Texts in Logic and Reasoning

Volume 1

Introduction to Deontic Logic and Normative Systems

Volume 1
Introduction to Deontic Logic and Normative Systems
Xavier Parent and Leendert van der Torre

Texts in Logic and Reasoning Series Editors
Leon van der Torre
Xavier Parent

Introduction to Deontic Logic and Normative Systems

Xavier Parent
Leendert van der Torre

ISBN 978-1-84890-269-5

College Publications
Scientific Director: Dov Gabbay
Managing Director: Jane Spurr

http://www.collegepublications.co.uk

Cover produced by Laraine Welch
Printed by Lightning Source, Milton Keynes, UK

Contents

List of Figures

Preface

Deontic logic deals with obligation, permission and related normative concepts. This textbook introduces three systems that have dominated the landscape of deontic logic: monadic deontic logic, dyadic deontic logic, and input/output logic. It describes their language, semantics, proof theory, and gives soundness and completeness theorems. The addition of exercises makes the book ideal for self-study or as a textbook in class.

This textbook remains neutral on application issues. Since its inception, deontic logic has been applied in a variety of fields, including philosophy, ethics, linguistics, computer science, and the law. This textbook will serve as a valuable resource for students and researchers wishing to gain a practical understanding of deontic logic for use in their work. We assume basic knowledge of propositional logic, its proof theory and model theory, but no more.

This textbook is intended as a companion to the *Handbook of Deontic Logic and Normative Systems* [12]. The handbook is broader in scope, and presents a detailed overview of the main lines of research on contemporary deontic logic and related topics. Besides discussing the three frameworks presented in this textbook, the handbook also gives more historical background on deontic logic, and provides an in-depth study of topics of central interest in deontic logic, like conflicts between obligations, permission and constitutive norm.

Monadic deontic logic and dyadic deontic logic are modal logic frameworks. They come with a possible worlds semantics and a Hilbert-style proof theory. Input/output (I/O) logic falls within the category of what has been called a "norm-based" system by Hansen [20]. In computer science and artificial intelligence, the name "rule-based" system is often used with the same or nearly the same meaning, and throughout this book we will use both names interchangeably. In a norm-based or a rule-based system, the semantics of the deontic concepts is given, not with reference to a set of possible worlds, but with reference to an explicitly given code or normative system. This one is defined as a set of if-then statements or rules. In the particular case of I/O logic, an if-then statement is represented as a pair of formulas. The semantics is (as we call it) "operational". It is given by a set of procedures yielding outputs for inputs. The proof theory resembles the one used for conditional logics. It is given by a set of inference rules acting not just on formulas but on pairs of formulas.

We have found it convenient to split the book into two corresponding parts. Part I deals with deontic logic viewed as a branch of modal logic. Part II deals with norm-based deontic logic, focusing on input/output logic.

This texbook is based on lectures that have been given by the authors at the University of Luxembourg over the past seven years. The authors acknowledge support from the European Union's Horizon 2020 research and innovation programme under the Marie Skłodowska-Curie grant agreement No 690974 (MIning and REasoning with Legal texts, MIREL).

Part I

Deontic Logic as a Branch of Modal Logic

Chapter 1

Monadic Deontic Logic

1.1 Introduction

The topic of this chapter is monadic deontic logic, a family of systems based on an analogy between the deontic operator "It is obligatory that" and the alethic modal operator "It is necessary that".

1.2 Language

Definition 1. *Let $\mathbb{P}$ be a set of atomic propositions. The language $\mathcal{L}$ of monadic deontic logic is generated by the following BNF (Backus Normal Form):*

$$\phi ::= p \mid \neg\phi \mid (\phi \wedge \phi) \mid \bigcirc\phi$$

where $p \in \mathbb{P}$.

   The construct $\bigcirc\phi$ is read as "It is obligatory that ϕ." Other connectives are introduced by the definitions:

disjunction	$\phi \vee \psi$	*is*	$\neg(\neg\phi \wedge \neg\psi)$
implication	$\phi \rightarrow \psi$	*is*	$(\neg\phi) \vee \psi$
equivalence	$\phi \leftrightarrow \psi$	*is*	$(\phi \rightarrow \psi) \wedge (\psi \rightarrow \phi)$
verum	$\top$	*is*	$p \vee \neg p$ *(for some $p \in \mathbb{P}$)*
falsum	$\bot$	*is*	$\neg\top$
permission	$P\phi$	*is*	$\neg \bigcirc \neg\phi$
prohibition	$F\phi$	*is*	$\bigcirc\neg\phi$

The above Backus-Naur Form (BNF) is a convenient abbreviation for the following alternative definition:

Definition 2. *The language of monadic deontic logic is the smallest set $\mathcal{L}$ such that:*
- $\mathbb{P} \subseteq \mathcal{L}$
- *If $\phi \in \mathcal{L}$, then $\neg\phi \in \mathcal{L}$ and $\bigcirc\phi \in \mathcal{L}$*
- *If $\phi \in \mathcal{L}$ and $\psi \in \mathcal{L}$, then $\phi \wedge \psi \in \mathcal{L}$*

We omit outermost parentheses if doing so does not lead to confusion.

Remark 1. The definition allows for "iterated" modalities like $\bigcirc \bigcirc p$. "Mixed" formulas like $p \wedge \bigcirc q$ are also allowed.

Remark 2. $\mathcal{L}$ is the object-language. The Greek letters ϕ, ψ ... are not part of it; they belong to the meta-language. They are meta-variables ranging over elements in $\mathcal{L}$, placeholders for object-language formulas of a given category. The convention is that a meta-variable is to be uniformly substituted with the same instance in all its occurrences in a given more complex formula. This one is called a schema.

1.3   Relational semantics

Definition 3 (Relational model). *A relational model M is a t-uple*

$$(W, R, V)$$

where:

- W is a (non-empty) set of states (also called "possible worlds") $s, t, \ldots W$ is called the universe of the model.
- $R \subseteq W \times W$ is a binary relation over W. It is understood as a relation of deontic alternativeness: sRt (or, alternatively, $(s,t) \in R$) says that t is an ideal alternative to s, or that t is a "good" successor of s. The first one is "good" in the sense that it complies with all the obligations true in the second one. Furthermore, R is subject to the following constraint:

$$(\forall s \in W)(\exists t \in W)(sRt) \qquad \text{(seriality)}$$

This means that the model does not have a dead end, a state with no good successor.
- $V : \mathbb{P} \mapsto 2^W$ is a valuation function assigning to each atom p a set $V(p) \subseteq W$ (intuitively the set of states at which p is true).

We can represent a relational model using a graph. This is illustrated with figure 1.1.

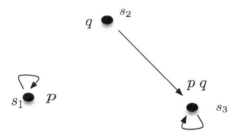

Figure 1.1: A 3-state model

States are nodes, and the content of the deontic alternativeness relation is indicated by arrows from nodes to nodes. For instance, $s_2 \to s_3$ says that s_3 is a good successor of s_2. In addition, each state s is labelled with the atoms made true at s. Here $\mathbb{P} = \{p, q\}$, and $V(p) = \{s_1, s_3\}$ while $V(q) = \{s_2, s_3\}$.

⚠ In the specification of a relational model M, if is not explicitly said that sRt, you should conclude that not-(sRt) is the case.

The satisfaction relation determines the truth-value of sentences according to their form.

Definition 4 (Satisfaction). *Given a relational model $M = (W, R, V)$ and a state $s \in W$, we define the satisfaction relation $M, s \vDash \phi$ (read as "state s satisfies ϕ in model M", or as " s makes ϕ true M") by induction on the structure of ϕ using the following clauses:*

- $M, s \vDash p$ *iff* $s \in V(p)$
- $M, s \vDash \neg\phi$ *iff* $M, s \nvDash \phi$
- $M, s \vDash \phi \wedge \psi$ *iff* $M, s \vDash \phi$ *and* $M, s \vDash \psi$
- $M, s \vDash \bigcirc\phi$ *iff for all* $t \in W$, *if* sRt *then* $M, t \vDash \phi$

Intuitively, $\bigcirc\phi$ is satisfied/true at s in model M just in case ϕ is satisfied/true at each of s'ideal alternatives.

We drop reference to M, and write $s \vDash \phi$, when it is clear what model is intended.

Example 1. Consider the model shown in figure 1.1. We have

- $s_1 \vDash p$ (since $s_1 \in V(p)$)
- $s_1 \vDash \neg q$ (since $s_1 \nvDash q$)
- $s_1 \vDash p \wedge \neg q$
- $s_1 \vDash \bigcirc(p \wedge \neg q)$ (s_1 has only one good successor: itself).

Validity, semantical consequence and satisfiability are always relative to a given class of models. In this sub-section they are stated with respect to the class of relational models as defined *supra*.

Definition 5 (Validity). *A formula ϕ is valid (notation: $\vDash \phi$) whenever, for all relational models $M = (W, R, V)$ and all states $s \in W$, $s \vDash \phi$.*

Example 2. It is of the nature of obligations that they are violable in principle. This is means that the deontic analogue of the so-called T axiom $\bigcirc\phi \rightarrow \phi$ is not valid. Consider the instance $\bigcirc p \rightarrow p$. Put $M = (W, R, V)$ with $W = \{s_1, s_2\}$, $s_1 R s_2$ and $V(p) = \{s_2\}$. Then, $s_1 \vDash \bigcirc p$ and $s_1 \nvDash p$. Thus, the schema $\bigcirc\phi \rightarrow \phi$ is not valid.

Definition 6 (Semantical consequence). *Given a set Γ of formulas, a formula ϕ is a semantical consequence of Γ (notation: $\Gamma \vDash \phi$) whenever, for all relational models $M = (W, R, V)$, and all $s \in W$, if $s \vDash \psi$ for all $\psi \in \Gamma$, then $s \vDash \phi$.*

We omit curly brackets for singleton sets for ease of reading.

⚠ There is no constraint on the cardinality of Γ. It can be infinite.

Example 3. $\bigcirc$ does not distribute over $\vee$. For instance, $\bigcirc p \vee \bigcirc q$ is not a semantical consequence of $\bigcirc(p \vee q)$. Figure 1.2 shows a relational model M and a state s in M, which satisfies $\bigcirc(p \vee q)$, but not $\bigcirc p \vee \bigcirc q$.

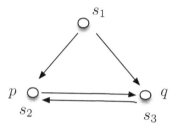

Figure 1.2: Failure of distribution of $\bigcirc$ over $\vee$

Definition 7 (Satisfiability). *A set Γ of formulas is satisfiable if and only if there is a relational model $M = (W, R, V)$, and some state $s \in W$ such that $s \vDash \phi$ for all $\phi \in \Gamma$.*

Remark 3. Satisfiability is the semantical counterpart of consistency.

Proposition 1. *We have*

$$\{\psi_1, ..., \psi_n\} \vDash \phi \text{ iff } \vDash (\psi_1 \wedge ... \wedge \psi_n) \to \phi \quad \text{(Deduction theorem)}$$
$$\text{If } \Gamma \vDash \phi \text{ and } \Gamma \subseteq \Gamma', \text{ then } \Gamma' \vDash \phi \quad \text{(Monotony)}$$

Proof. This follows at once from the definitions. ☐

1.4 Proof system

1.4.1 System D

A proof or a derivation in a given proof system, call it **X**, is a finite sequence of formulas each of which is an instance of an axiom schema or follows from earlier formulas in the sequence by a rule schema. If there is a derivation for ϕ in **X** we write $\vdash_X \phi$, or, if the system **X** is clear from the context, we just write $\vdash \phi$. We then also say that ϕ is a *theorem* in **X**, or that **X** proves ϕ.

Definition 8. D *is the proof system consisting of the following axiom schemas and rule schemas.*

$\vdash \phi$, *where ϕ is a propositional tautology*	(PL)
$\vdash \bigcirc(\phi \to \psi) \to (\bigcirc\phi \to \bigcirc\psi)$	($\bigcirc$-K)
$\vdash \bigcirc\phi \to P\phi$	($\bigcirc$-D)
If $\vdash \phi$ and $\vdash \phi \to \psi$ then $\vdash \psi$	(MP)
If $\vdash \phi$ then $\vdash \bigcirc\phi$	($\bigcirc$-Nec)

⚠ Strictly speaking, instead of PL, we should have listed the axioms for propositional logic. We use PL, as is common in the axiomatic presentations of modal systems, because it helps to simplify the proofs.

Remark 4. $\bigcirc$-K and $\bigcirc$-D are the deontic analogues of the modal axioms K and D. $\bigcirc$-D can equivalently be written as $\vdash \neg(\bigcirc\phi \wedge \bigcirc\neg\phi)$. It rules out the possibility of conflicts between obligations. $\bigcirc$-Nec is the deontic analogue of the rule of necessitation.

Remark 5. Do not confuse $\bigcirc$-Nec with the law $\vdash \phi \to \bigcirc\phi$. $\bigcirc$-Nec takes the form of a conditional statement, whose antecedent and consequent are theorems. Recast in model-theoretic terms, $\bigcirc$-Nec states that, if $\models \phi$ then $\models \bigcirc\phi$. This is obviously not the same as $\models \phi \to \bigcirc\phi$.

Example 4. Below: a derivation of $\bigcirc(p \wedge q) \to \bigcirc p$.

1. $\vdash ((p \wedge q) \to p)$	PL
2. $\vdash \bigcirc((p \wedge q) \to p)$	$\bigcirc$-Nec, 1
3. $\vdash \bigcirc((p \wedge q) \to p) \to (\bigcirc(p \wedge q) \to \bigcirc p)$	$\bigcirc$-K
4. $\vdash \bigcirc(p \wedge q) \to \bigcirc p$	MP, 2,3

Definition 9 (Syntactical consequence). *Given a set* Γ *of formulas, we say that* ϕ *is a syntactical consequence of* Γ *(notation:* $\Gamma \vdash \phi$*) iff there are formulas* $\psi_1, \ldots, \psi_n \in \Gamma$ *such that* $\vdash (\psi_1 \wedge \ldots \wedge \psi_n) \to \phi$*. (In case where* $n = 0$*, this means that* $\vdash \phi$*.)*

Definition 10 (Consistency). *A set* Γ *is consistent if* $\Gamma \nvdash \bot$*, and inconsistent otherwise.*

1.4.2   Soundness and completeness theorem

The role of a soundness and completeness theorem is to establish that the semantic and proof-theoretic characterisation of the logic are equivalent.

Theorem 1 (Soundness, weak version). *If* $\vdash \phi$ *then* $\vDash \phi$.

Proof. It is enough to show that the axiom schemas are valid in the class of relational models, and that the rules schema preserve validity in the class of relational models.

We give the argument for $\bigcirc$-K. Consider some M and some s such that $s \vDash \bigcirc(\phi \to \psi)$ and $s \vDash \bigcirc\phi$. For the result that $s \vDash \bigcirc\psi$, we need to show that

For all t in M, if sRt, then $t \vDash \psi$.

So let t in M be such that sRt. By the two opening assumptions, one gets $t \vDash \phi \to \psi$ and $t \vDash \phi$. This follows from the truth-conditions for $\bigcirc$. By the truth-conditions for $\to$, one then gets $t \vDash \psi$ as required. $\square$

Theorem 2 (Soundness, strong version). *If* $\Gamma \vdash \phi$*, then* $\Gamma \vDash \phi$.

Proof. Let $\Gamma \vdash \phi$. By definition, there are $\psi_1, \ldots, \psi_n \in \Gamma$ such that $\vdash (\psi_1 \wedge \ldots \wedge \psi_n) \to \phi$. By weak soundness, $\vDash (\psi_1 \wedge \ldots \wedge \psi_n) \to \phi$. By the deduction theorem, Proposition 1, $\{\psi_1, ..., \psi_n\} \vDash \phi$. By monotony, Proposition 1 again, $\Gamma \vDash \phi$. $\square$

Theorem 3 (Completeness, weak and strong version). *i) If* $\vDash \phi$ *then* $\vdash \phi$*; ii) if* $\Gamma \vDash \phi$ *then* $\Gamma \vdash \phi$.

Proof. Weak completeness follows from strong completeness, which in turn can be established using the method of canonical models. For details, see Chellas [9] or Åqvist [4]. □

Remark 6. The difference between the weak and strong versions of the soundness and completeness theorem is that with the strong version there is no restriction on the cardinality of Γ. This one may be (countably) infinite.

As a spin-off of theorem 3, one gets:

Corollary 1 (Semantic compactness). *A set Γ of formulas is satisfiable, if every finite Δ ⊆ Γ is satisfiable.*

1.4.3   Decidability

A model M is called finite, if its universe W has finitely many elements.

Theorem 4 (Finite model property). *If ϕ is satisfiable in a relational model, then ϕ is satisfiable in a finite relational model.*

Proof. Using the so-called filtration method. Details can be found in Chellas [9]. □

The theoremhood problem in a given system is decidable if there is a decision procedure–an effective finitary method–for determining whether an arbitrary formula is a theorem.

Theorem 5 (Decidability). *The theoremhood problem in **D** is decidable.*

Proof. This follows from the finite model property and the fact that **D** is a finitely axiomatized logic in the usual way. Roughly speaking the reason is as follows. The fact that ϕ is a theorem can be verified in finitely many steps, because a proof is always finite. The fact that ϕ is not a theorem can also be verified in finitely many steps. For if it is not a theorem, then after examining a finite number of models one will appear that satisfies $\neg\phi$. □

1.5   Stronger systems

This section describes three extensions of **D** that have been considered in the literature: **DS4** (Deontic S4), **DS5** (Deontic S5) and **DM** (Deontic M).

DS4 extends **D** with the axiom schema

$$\bigcirc \phi \rightarrow \bigcirc \bigcirc \phi \tag{$\bigcirc$-4}$$

DS4 is sound and complete with respect to the class of relational models in which R is serial and transitive:

$$(\forall s)(\forall t)(\forall u)(sRt \;\&\; tRu \rightarrow sRu) \tag{transitivity}$$

DS5 extends **DS4** with the axiom schema

$$\neg \bigcirc \phi \rightarrow \bigcirc \neg \bigcirc \phi \tag{$\bigcirc$-5}$$

DS5 is sound and complete with respect to the class of relational models in which R is serial, transitive and euclidean:

$$(\forall s)(\forall t)(\forall u)(sRt \;\&\; sRu \rightarrow tRu) \tag{euclideanness}$$

DM is obtained by replacing, in **DS5**, $\bigcirc$-5 with the axiom schema

$$\bigcirc (\bigcirc \phi \rightarrow \phi) \tag{$\bigcirc$-M}$$

DM is sound and complete with respect to the class of relational models in which R is serial, transitive and secondary reflexive:

$$(\forall s)(\forall t)(sRt \rightarrow tRt) \tag{secondary reflexivity}$$

Note that $\bigcirc$-M is a theorem of **DS5**. Hence these systems form a series of systems of increasing strength:

$$\textbf{D} \subset \textbf{DS4} \subset \textbf{DM} \subset \textbf{DS5}$$

Theorem 6 (Completeness, strong version). *DS4, DM and DS5 are sound and strongly complete with respect to the class of relational models in which R meets the required conditions as stated.*

Theorem 7 (Decidability). *The theoremhood problem in DS4, DM and DS5 is decidable.*

1.6 Chisholm's paradox

Chisholm [10]'s paradox refers to the problem of reasoning about norm violation. More specifically, it refers to a problem raised by the formalisation of the following four sentences:

(A) It ought to be that Jones goes to the assistance of his neighbours;

(B) It ought to be that, if Jones goes to the assistance of his neighbours, then he tells them he is coming;

(C) If Jones does not go to the assistance of his neighbours, then he ought not to tell them he is coming;

(D) Jones does not go to the assistance of his neighbours.

(A) expresses what is usual called a primary obligation. (B) expresses what is called an according-to-duty (ATD) obligation: it says what should be done if the primary obligation is fulfilled. (C) expresses what is called a contrary-to-duty (CTD) obligation: it says what should be done if the primary obligation is violated. (D) is a fact. It tells us that the primary obligation is violated.

The problem is this. Intuitively, the sentences (A)-(D) are mutually consistent, and they are logically independent one from the other. Their formalisation in $\mathbf{D}$ either makes them mutually inconsistent or makes one sentence derivable from another one in this set. This is Chisholm's paradox.

There are four candidate formalisations in $\mathbf{D}$, depending on whether $\bigcirc$ takes wide or narrow scope over $\to$. They are listed below:

(A_0) $\bigcirc g$ (A_1) $\bigcirc g$
(B_0) $g \to \bigcirc t$ (B_1) $\bigcirc(g \to t)$
(C_0) $\bigcirc(\neg g \to \neg t)$ (C_1) $\neg g \to \bigcirc \neg t$
(D_0) $\neg g$ (D_1) $\neg g$

(A_2) $\bigcirc g$ (A_3) $\bigcirc g$
(B_2) $\bigcirc(g \to t)$ (B_3) $g \to \bigcirc t$
(C_2) $\bigcirc(\neg g \to \neg t)$ (C_3) $\neg g \to \bigcirc \neg t$
(D_2) $\neg g$ (D_3) $\neg g$

(A_0)-(D_0) is ruled out from the outset, because we would derive neither $\bigcirc t$ nor $\bigcirc \neg t$ from it. None of the other candidate formalisations meets both the requirement of consistency and that of logical independence. For instance,

Proposition 2. *The set* $\Gamma = \{A_1, B_1, C_1, D_1\}$ *is not satisfiable, and thus it is inconsistent.*

Proof. Assume there exist a relational model $M = (W, R, V)$ and a state $s_1 \in W$ such that

- $M, s_1 \vDash \bigcirc g$
- $M, s_1 \vDash \bigcirc (g \to t)$
- $M, s_1 \vDash \neg g \to \bigcirc \neg t$
- $M, s_1 \vDash \neg g$.

From $M, s_1 \vDash \neg g$ and $M, s_1 \vDash \neg g \to \bigcirc \neg t$ we deduce $M, s_1 \vDash \bigcirc \neg t$. By seriality we know there is $s_2 \in W$ such that $s_1 R s_2$. By the definition of $\bigcirc$ we know $M, s_2 \vDash \neg t$, $M, s_2 \vDash g$, $M, s_2 \vDash g \to t$, which is a contradiction.

The above shows that Γ is not satisfiable, definition 7. $\qquad \square$

1.7 The "Andersonian" reduction

This section shows how to embed monadic deontic logic into alethic modal logic. The embedding is based on Anderson [2]'s suggestion that a statement like "it ought to be that ϕ" can be analysed as "if $\neg\phi$, then necessarily v", where v is a propositional constant expressing that there has been a violation of the norms, or that a bad state-of-affairs has occurred.

For technical reasons, it is best to work with a language that does not distinguish between primitive and derived connectives. This is what Åqvist [4, 5] does. Thus, in this section, we assume that $\vee, \to, \leftrightarrow, \top, \bot$ and P are all part of the alphabet of the language of **D**, on a par with $\neg, \wedge$ and $\bigcirc$. The axiomatic characterization of **D** is amended accordingly in a straightforward way, by adding as an axiom the definitions that were used to introduce the derived connectives. For instance, **D** contains $P\phi \leftrightarrow \neg \bigcirc \neg \phi$ as an extra axiom. We do the same for the system of alethic modal logic into which **D** is embedded.

1.7.1 System $\mathbf{K}^v$

Definition 11. *The language of* $\mathbf{K}^v$ *is generated by the following BNF:*

$$\phi ::= p \mid v \mid \top \mid \bot \mid \neg\phi \mid \phi \vee \phi \mid \phi \wedge \phi \mid \phi \rightarrow \phi \mid \phi \leftrightarrow \phi \mid \Box\phi \mid \Diamond\phi$$

where $p \in \mathbb{P}$. v *is a designated propositional constant read as "a viola-tion has occurred".* $\Box\phi$ *is read as "It is necessary that* ϕ*".* $\Diamond\phi$ *is read as "It is possible that* ϕ*".*

Definition 12. *An Anderson model is a tuple* $M = (W, S, B, V)$ *such that*

- W *is a (non-empty) set of states*
- $S \subseteq W \times W$ *is a binary relation over* W. *It is understood as a relation of alethic accessibility:* sSt *(or, alternatively,* $(s, t) \in S$*) says that* t *is accessible from* s, *or that* t *is a successor of* s.
- $B \subseteq W$ *is a set of "bad" states. It is required that,*

$$(\forall s \in W)(\exists t \in W)(sSt \,\&\, t \notin B) \qquad (\$)$$

Intuitively: for every state, there is an accessible one that is good; every state has a good successor.

- V *is as before.*

⚠ Do not confuse S with the relation R used for system $\mathbf{D}$

Definition 13 (Satisfaction). *Given an Anderson model* M *and a state* $s \in W$, *we define the satisfaction relation* $M, s \vDash \phi$ *using the same clauses as before plus:*

- $M, s \vDash v$ *iff* $s \in B$
- $M, s \vDash \Box\phi$ *iff, for all* $t \in W$, *if* sSt *then* $M, t \vDash \phi$
- $M, s \vDash \Diamond\phi$ *iff, for some* $t \in W$, sSt *and* $M, t \vDash \phi$

Definition 14. $\mathbf{K}^v$ *is the proof system obtained by supplementing so-*

called system **K** *(after Kripke) with the axiom schema AV.*

$\vdash \phi$, *where ϕ is a propositional tautology*	(PL)
$\vdash \Box(\phi \rightarrow \psi) \rightarrow (\Box\phi \rightarrow \Box\psi)$	($\Box$-K)
$\vdash \Diamond\phi \leftrightarrow \neg\Box\neg\phi$	($\Box$-D)
If $\vdash \phi$ *and* $\vdash \phi \rightarrow \psi$ *then* $\vdash \psi$	(MP)
If $\vdash \phi$ *then* $\vdash \Box\phi$	($\Box$-Nec)
$\vdash \Diamond\neg v$	(AV)

PL, $\Box$-K, $\Box$-D, MP and $\Box$-Nec are the axioms of **K**. *Intuitively, AV says that the bad thing, v, is avoidable.*

Theorem 8 (Soundness and completeness, strong version). *System* $\mathbf{K}^v$ *is strongly sound and strongly complete with respect to the class of Anderson models.*

Proof. Soundness is a matter of showing that the axiom schemas of $\mathbf{K}^v$ are valid in the class of Anderson models, and that the rule schemas preserve validity in the class of Anderson models. We just show AV. Let M be an Anderson model, and s be a state in M. By condition (\$) in definition 12, there is some t such that sSt and $t \notin B$. By the truth-conditions for v, $t \not\models v$, and so $t \models \neg v$. By the truth-conditions for $\Diamond$, $s \models \Diamond\neg v$. Hence, $\models \Diamond\neg v$.

Completeness is shown using canonical models. $\Box$

Theorem 9 (Decidability). *The theoremhood problem in* $\mathbf{K}^v$ *is decidable.*

1.7.2   Embedding D into $\mathbf{K}^v$

We recall the basic idea underpinning an embedding. We have two logics L_1 (source logic) and L_2 (target logic). The first step consists in showing how to translate a formula ϕ in the language of L_1 into a formula $\tau(\phi)$ in the language of L_2. The second step consists in showing that the embedding is faithful: the original formula ϕ is derivable in the source logic L_1 iff its translation $\tau(\phi)$ is derivable in the target logic L_2.

Definition 15. *An embedding of* **D** *into* $\mathbf{K}^v$ *is a function* τ *mapping each formula* ϕ *in the language of* **D** *into a formula* $\tau(\phi)$ *in the language of* $\mathbf{K}^v$. τ *is defined inductively as follows:*

$$\tau(p) = p \tag{1.1}$$
$$\tau(\bot) = \bot \tag{1.2}$$
$$\tau(\top) = \top \tag{1.3}$$
$$\tau(\neg\phi) = \neg\tau(\phi) \tag{1.4}$$
$$\tau(\phi \wedge \psi) = \tau(\phi) \wedge \tau(\psi) \tag{1.5}$$
$$\tau(\phi \vee \psi) = \tau(\phi) \vee \tau(\psi) \tag{1.6}$$
$$\tau(\phi \to \psi) = \tau(\phi) \to \tau(\psi) \tag{1.7}$$
$$\tau(\phi \leftrightarrow \psi) = \tau(\phi) \leftrightarrow \tau(\psi) \tag{1.8}$$
$$\tau(\bigcirc\phi) = \Box(\neg\tau(\phi) \to v) \tag{1.9}$$
$$\tau(P\phi) = \Diamond(\tau(\phi) \wedge \neg v) \tag{1.10}$$

Example 5.

$$\tau(\bigcirc p \to Pp) = \tau(\bigcirc p) \to \tau(Pp)$$
$$= \Box(\neg\tau(p) \to v) \to \Diamond(\tau(p) \wedge \neg v)$$
$$= \Box(\neg p \to v) \to \Diamond(p \wedge \neg v)$$

Theorem 10 (Faithfulness of the embedding, LTR). *For every formula* ϕ *in the language of* **D**, *if* $\vdash_\mathbf{D} \phi$, *then* $\vdash_{\mathbf{K}^v} \tau(\phi)$.

Proof. By induction on the length of a proof. $\qquad\Box$

Lemma 1. *For every relational model* $M = (W, R, V)$, *there is an Anderson model* $M' = (W, S, B, V)$ *(with* W *and* V *the same) such that:*

$$\text{For all states } s, M, s \models \phi \text{ iff } M', s \models \tau(\phi) \tag{1.11}$$

Proof. Put $S = R$ and $B = \{s : s \in W \ \& \ \neg\exists t\, tRs\}$. It follows that

$$sRt \text{ iff: } sSt \ \& \ t \notin B \tag{1.12}$$

Because R is serial, S and B meet the property required of them in definition 12. The proof of (1.11) is by induction on the structure of ϕ. $\qquad\Box$

Theorem 11 (Faithfulness of the embedding, RTL). *For every formula* ϕ *in the language of* **D**, *if* $\vdash_{\mathbf{K}^v} \tau(\phi)$, *then* $\vdash_{\mathbf{D}} \phi$.

Proof. We show the contrapositive. Suppose $\nvdash_{\mathbf{D}} \phi$. By the completeness theorem for **D**, there exists a relational model $M = (W, R, V)$, and a state s in M, such that $M, s \nvDash \phi$. By Lemma 1, there exists an Anderson model $M' = (W, S, B, V)$, with s in M', and such that $M', s \nvDash \tau(\phi)$. By the soundness theorem for $\mathbf{K}^v$, $\nvdash_{\mathbf{K}^v} \tau(\phi)$. □

1.7.3 Chisholm's scenario revisited

In the modal logic of C. I. Lewis, if it is necessary that ϕ implies ψ (notation: $\square(\phi \rightarrow \psi)$), then ϕ is said to strictly imply ψ (notation: $\phi \prec \psi$). Anderson's reduction hints at a possible formalisation of Chisholm's scenario using strict implication $\prec$ in place of material implication:

(A$_4$) $\bigcirc g$
(B$_4$) $g \prec \bigcirc t$
(C$_4$) $\neg g \prec \bigcirc \neg t$
(D$_4$) $\neg g$

The formula (A$_4$) here is a shorthand for $\neg g \prec v$, viz. $\square(\neg g \rightarrow v)$. The same applies to the formulas prefixed with $\bigcirc$ appearing in (B$_4$) and (C$_4$). For instance, written in full, (B$_4$) is $\square(g \rightarrow \square(\neg t \rightarrow v))$.

Proposition 3. *The set* $\Gamma = \{A_4, B_4, C_4, D_4\}$ *is satisfiable in an Anderson model, and thus it is consistent in* $\mathbf{K}^v$.

Proof. Put $M = (W, S, B, V)$ with $W = \{s_1, s_2, s_3\}$, $s_1 S s_1$, $s_1 S s_2$, $s_2 S s_3$, $s_3 S s_3$, $B = \{s_1\}$, $V(g) = \{s_2, s_3\}$, $V(t) = \{s_1, s_3\}$ and $V(v) = \{s_1\}$. This is illustrated with figure 1.3, where the content of the alethic accessibility relation S is indicated by dashed arrows. The property required of S and B in definition 12 is met. We have:

- $s_1 \vDash \bigcirc g$
- $s_1 \vDash \square(g \rightarrow \bigcirc t)$
- $s_1 \vDash \square(\neg g \rightarrow \bigcirc \neg t)$
- $s_1 \vDash \neg g$

□

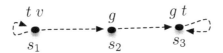

Figure 1.3: Chisholm's example–Anderson

1.8   Neighbourhood/minimal semantics

The neighbourhood/minimal semantics [9] is a well-known generalisation of the relational semantics. It has been used in deontic logic too.

Definition 16 (Minimal model). *A minimal model $M = (W, N, V)$ is a structure where W and V are as before, and N, called a "neighbourhood" function, is a function assigning to each state $s \in W$ a set of subsets of W (i.e. $N(s) \subseteq 2^W$ for each $s \in W$).*

Definition 17 (Satisfaction). *Given a minimal model $M = (W, N, V)$, and a world $s \in W$, we define the satisfaction relation $M, s \vDash \phi$ as before, except for deontic formulas, where:*

$$M, s \vDash \bigcirc \phi \ \textit{iff} \ \|\phi\| \in N(s)$$

Here $\|\phi\|$ is the truth-set of ϕ, viz the set $\{t \in W : M, t \vDash \phi\}$. $\|\phi\|$ may be read as the proposition expressed by ϕ. Intuitively, for each s, $N(s)$ is a collection of propositions that are standards of obligation relative to s, and $\bigcirc \phi$ is true at s just in case the proposition expressed by ϕ is entailed by one of these standards.

Validity and consequence are defined as in the relational semantics. The neighbourhood approach gives an extra degree of freedom. The obligation operator as defined in definition 17 is very weak. But extra constraints may be placed on $N(s)$ in order to make the operator validate more laws, as one thinks fit. It is known that, when sufficiently strong constraints are placed on $N(s)$, one obtains the same logic as with the relational semantics.

For an illustration, consider the minimal deontic logic proposed by Chellas [9]. It has the law OIC, but not the law $\bigcirc$-D:

$$\neg \bigcirc \bot \qquad \qquad \text{(OIC)}$$
$$\neg(\bigcirc\phi \wedge \bigcirc\neg\phi) \qquad \qquad \text{($\bigcirc$-D)}$$

OIC expresses the seemingly uncontroversial principle *"ought* implies *can"*. $\bigcirc$-D rules out the possibility of conflicts between obligations. This seems to be counter-intuitive. For in real-life, conflicts between obligations are common-place. In **D** and stronger systems, we have $\vdash (\neg\bigcirc\bot) \leftrightarrow \neg(\bigcirc\phi\wedge\bigcirc\neg\phi))$. This makes it impossible to distinguish between OIC and $\bigcirc$-D, and to have one without the other. The proof theory of Chellas's minimal deontic logic is given by

$\vdash \phi$, where ϕ is a propositional tautology (MP)
$\neg \bigcirc \bot$ (OIC)
If $\vdash \phi \to \psi$ then $\vdash \bigcirc\phi \to \bigcirc\psi$ (ROM)

The relevant constraints to be placed on N are:

$$\emptyset \notin N(s)$$
$$\text{If } U \subseteq V \text{ and } U \in N(s) \text{ then } V \in N(s) \qquad \text{(closure under superset)}$$

Theorem 12. *Chellas's minimal deontic logic is sound and weakly complete with respect to the class of finite minimal models in which N meets the above two constraints.*

Proof. The proof is an adaptation of the proof of weak completeness of Chellas's system EK given by Goble [17]. □

As a spin-off result, one gets:

Theorem 13. *The theoremhood problem in Chellas's minimal deontic logic is decidable.*

Proof. Immediate from the above. □

1.9 Notes

Von Wright [54] is usually credited for having introduced the monadic modal approach to deontic logic [54]. His system became known as "standard deontic logic" (SDL), though this may be seen as a misnomer. SDL was a landmark system up to the late 60s only, when so-called Dyadic Deontic Logic (Chapter 2) will be devoted to it) emerged as a new standard. For some reason, the name SDL has stuck. Von Wright's system was quickly associated with the system **D** described in subsection 1.4.1, in spite of obvious differences between the two. **D** itself is no more nor less than the well-known modal system of type KD, where $\Box$ is read as an obligation operator.

The name **D** used for the system described in subsection 1.4.1 is taken from Hanson [21] and van Fraassen [53]. Other names have been used in the literature. For instance, **D** is called **D*** by Chellas [9] and **OK**$^+$ by Åqvist [5]. The names **DS4**, **DS5** and **DM** used for the systems described in subsection 1.5 are Hanson [21]'s. They are derived from the alethic modal logics **M**, **S4** and **S5**, to which they are similar. Other names have been used in the literature too. Hanson [21] applied the so-called method of tableaux to these systems.

For more on the historical background, the reader is referred to Hilpinen and McNamara [23].

One final remark on Anderson's reduction. The embedding described in subsection 1.7.2 is Åqvist [4, 5]'s. The only difference is that Åqvist works with Kanger's reduction schema, in terms of a propositional constant Q interpreted as "what morality prescribes". It is known that, if Q is defined as $\neg v$, then the two approaches are equivalent.

Anderson does not formulate the reduction the way we do, nor does he even give a semantics to his system. He formulates his reduction as we do in exercise 11. One might call this the "definitional" approach. Anderson's own formulation is simpler, but it has a drawback: it does not provide a guarantee that the encoding is faithful. To be more precise, it does not say anything about the status of the non-theorems of **D**. It could be that the alethic counterpart of some are theorems of the alethic modal logic. In [3], Anderson proposed that his reduction be worked out in the framework of the so-called logic **R** of relevant implication. This proposal is examined in Goble [13].

1.10 Exercises

Exercise 1 (section 1.2). Symbolize the following English statements:
1. You should pay tax;
2. You must wash your dish or put it in the washing machine;
3. It ought to be the case that whatever ought to be the case be the case.

Exercise 2 (section 1.3).
1. Define the truth-conditions for $\vee$ and $\rightarrow$ in a relational model;
2. Same question for P and F.

Exercise 3 (section 1.3). Consider the relational model depicted below.
1. Does the model have a dead end, a world with no good successor?
2. Is it, or is it not, the case that $s_1 \models \bigcirc Pp$? And is it, or is it not the case that $s_2 \models \bigcirc Pp$?
3. Which state satisfy the formula $\bigcirc p \rightarrow p$?

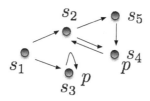

Exercise 4 (section 1.3). Explain in what sense the notion of validity may be described as a limiting case of the notion of consequence.

Exercise 5 (section 1.3). Show that $\bigcirc(\phi \vee \psi)$ is a semantical consequence of $\bigcirc\phi \vee \bigcirc\psi$, viz. $\bigcirc\phi \vee \bigcirc\psi \models \bigcirc(\phi \vee \psi)$.

Exercise 6 (section 1.4). Give the proof of proposition 1.

Exercise 7 (section 1.4).
1. Complete the proof of soundness, theorem 1;
2. Describe a model in which R is not serial, and in which the D axiom is not valid.

Exercise 8 (section 1.4). To show that a set Γ of formulas is inconsistent, we usually turn to the semantics, and show that Γ is not satisfiable. Explain why unsatisfiability shows inconsistency. (Hint: use the soundness and completeness theorem.)

Exercise 9 (section 1.5). Show that $\bigcirc$-M is a theorem of **DS5**.

Exercise 10 (section 1.6). Show that A_2-D_2 and A_3-D_3 do not meet the requirement of logical independence. Show that A_1-D_1 meets the requirement of logical independence.

Exercise 11 (section 1.7). Suppose Anderson's reduction schema is given the form of a pair of definitions (rather than that of an embedding function): $\bigcirc\phi =_{def} \Box(\neg\phi \to v)$; $P\phi =_{def} \Diamond(\phi \wedge \neg v)$. Show that all the axiom and rule schemas of **D** are derivable in $\mathbf{K}^v$.

Exercise 12 (section 1.7). Look at the Anderson model depicted in the diagram used in the proof of proposition 3.
1. Explain why the property required of S and B in definition 12 is met. Explain why we have that each of A_4, B_4, C_4 and D_4 is true at state s_1.
2. Give the relational model from which this Anderson model could have been derived. Hint: use backwards the construction given in the proof of lemma 1.
3. What do you conclude about the solution to Chisholm's paradox using Anderson's reduction schema?

Exercise 13 (section 1.8).
1. Show that ROM preserves validity in the class of minimal models meeting the property of closure under superset.
2. Define the evaluation rules for P and F in a minimal model.

Chapter 2

Dyadic Deontic Logic

2.1 Introduction

Dyadic Deontic Logic (DDL) can be described as a generalisation of MDL. On the syntactical side, one works with a primitive dyadic obligation operator $\bigcirc(-/-)$. It is meant to represent a notion of conditional obligation that is "weaker" than the one captured in MDL using material implication or strict implication. On the semantical side, one allows for grades of ideality in the models.

2.2 Language

Definition 18. *The language of DDL is generated by the following BNF:*

$$\phi ::= p \mid \neg\phi \mid \phi \wedge \phi \mid \Box\phi \mid \bigcirc(\phi/\phi)$$

$\Box\phi$ is read as "ϕ is settled as true", and $\bigcirc(\psi/\phi)$ as "ψ is obligatory, given ϕ". ϕ is called the antecedent, and ψ the consequent.

$P(\psi/\phi)$ ("ψ is permitted, given ϕ") is short for $\neg \bigcirc (\neg\psi/\phi)$, $\bigcirc\phi$ ("ϕ is unconditionally obligatory") and $P\phi$ ("ϕ is unconditionally permitted") are short for $\bigcirc(\phi/\top)$ and $P(\phi/\top)$, respectively. $\Diamond\phi$ is short for $\neg\Box\neg\phi$. Other Boolean connectives are defined as in the previous chapter.

23

Remark 7. Mixed formulas like $\bigcirc(q/p) \wedge p$ are allowed. Iterations of deontic modalities are allowed. For example, $\bigcirc(q/(\bigcirc(q/p) \wedge p))$ is a well-formed formula.

Example 6. A primary obligation is represented as $\bigcirc p$. Example: one ought to pay one's taxes. A contrary-to-duty (CTD) obligation (which says what should be done if a primary obligation is violated) has the form $\bigcirc(q/\neg p)$. Example: if one does not pay one's taxes, then one ought to pay a fine. An according-to-duty (ATD) obligation (which says what should be done if a primary obligation is fulfilled) has the form $\bigcirc(\neg r/p)$. Example: if one pays one's taxes, then one ought not to pay a fine.

2.3 Preference models

2.3.1 Basic setting

Definition 19. *A preference model* $M = (W, \succeq, V)$ *is a tuple where:*
- *W and V are as before.*
- *$\succeq$ is a binary relation over W ordering the states according to their relative goodness. $s \succeq t$ is read as "state s is at least as good as state t".*

Remark 8. Given a preference model $M = (W, \succeq, V)$, if is not explicitly said that $s \succeq t$, you should conclude that $s \not\succeq t$ is the case.

A state in which ϕ holds is called a ϕ-state. Intuitively, the evaluation rule of the dyadic operator puts $\bigcirc(\psi/\phi)$ true, whenever all the best ϕ-states are ψ-states. The formal definition of "best" is postponed until section 2.3.4.

Definition 20 (Satisfaction). *Given a preference model* $M = (W, \succeq, V)$ *and a state* $s \in W$, *we define the satisfaction relation* $M, s \vDash \phi$ *as before except for the following two new clauses*
- *$M, s \vDash \Box\phi$ iff, for all $t \in W$, $M, t \vDash \phi$*
- *$M, s \vDash \bigcirc(\psi/\phi)$ iff $best(\|\phi\|) \subseteq \|\psi\|$*

where $\|\phi\|$ *is the truth-set of* ϕ, *viz. the set of all the ϕ-states, and* $best(\|\phi\|)$ *the subset of those that are best according to* $\succeq$.

2.3.2 Meaning of the betterness relation

The betterness relation can be understood as an abstract relation of closeness to (full) ideality. This is to be contrasted with the binary classification of states into good/bad in monadic deontic logic. The idea is to allow for gradations between these two extremes. The closer to ideality a state is, the better it is. In the case of CTDs, one gets some reasonable spelling out of the betterness relation, by ranking states based on the number of obligations they violate: the more obligations are violated in a given state, the farther to ideality this state is. Here is a quote from literature:

> "We might interpret the 'betterness' relation between possible worlds by putting one world better than another iff it violates only a subset of the explicit obligations of the [normative] system, that are violated by the other." [31]

This can be illustrated with Chisholm's paradox. A state s is said to violate $\bigcirc(\psi/\phi)$ if $s \models \phi$ but $s \not\models \psi$.

Example 7 (Chisholm). Chisholm's scenario is formalized in DDL as

$$\Gamma = \{\bigcirc g, \bigcirc(t/g), \bigcirc(\neg t/\neg g), \neg g\}$$

Figure 2.1 describes a typical preference model of Γ. As before the convention is that at each point, viz. at each element $s \in W$, we list the elementary letters that s satisfies, omitting those that it makes false. s_1 is best, because it does not violate any obligation in Γ. s_2 and s_3 are second best, because they each violate one obligation. s_4 is worst, because it violates two obligations.

Figure 2.1 makes it clear why a binary classification of states into good/bad is not suitable to model CTDs. A state in which a primary obligation is violated, but the associated CTD obligation fulfilled, is better than a state in which both are violated. Yet, it is not as good as one in which none of the two obligations is violated.

Of course, other readings of $\succeq$ are possible. For instance, in some frameworks, like deontic STIT logic (cf. Horty [24]), its role is to encode utilities. Reference is then made to so-called utilitarianism, viz.

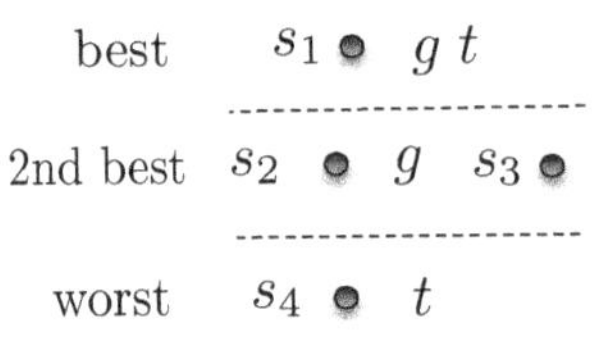

best s_1 ● $g\ t$
$\dashrightarrow$
2nd best s_2 ● g s_3 ●
$\dashrightarrow$
worst s_4 ● t

Figure 2.1: A preference model of the Chisholm set

the theory in normative ethics holding that the best moral action is the one that maximizes utility.

We now give more details on some of the components of the semantics.

2.3.3   Some more value "relations"

Definition 21 (Equal goodness). *Two states s and t are equally good (notation: $s \simeq t$) if $s \succeq t$ and $t \succeq s$.*

Definition 22 (Incomparability). *Two states s and t are incomparable (notation: $s\|t$) if $s \not\succeq t$ and $t \not\succeq s$.*

Definition 23 (Strict betterness). *The strict relation $\succ$ induced by $\succeq$ is defined by: $s \succ t$ iff $s \succeq t$ and $t \not\succeq s$. For $s \succ t$, read "s is strictly better than t".*

2.3.4   Best as optimal

Intuitively, the evaluation rule for the conditional obligation operator says that $\bigcirc(\psi/\phi)$ holds iff ψ holds in all the best states that make ϕ true. It remains to give the formal definition of "best". Here we use the following definition–it is by no means the only possible one.

Definition 24 (Optimality). *For state s to qualify as best among the ϕ-states, s must be optimal under $\succeq$: s makes ϕ true, and s is at least as good as any other states t making ϕ true. Formally:*

$$\text{best}_{\succeq}(\|\phi\|) = \text{opt}_{\succeq}(\|\phi\|)$$
$$= \{s \in \|\phi\| \mid \forall t \ (t \vDash \phi \rightarrow s \succeq t)\}$$

⚠ Keep in mind that, for s to qualify as a best ϕ-state, it must satisfy ϕ in the first place.

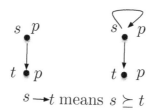

$s \rightarrow t$ means $s \succeq t$

Figure 2.2: Best as optimal

Example 8. Consider figure 2.2, where $s \rightarrow t$ represents $s \succeq t$. In the diagram to the left, $\text{best}_{\succeq}(\|p\|) = \emptyset$ because $s \not\succeq s$, while in the diagram to the right $\text{best}_{\succeq}(\|p\|) = \{s\}$ because $s \succeq s$.

⚠ The inclusion $\text{opt}_{\succeq}(\|\phi\|) \subseteq \|\psi\|$ will be referred to as the opt-rule.

2.3.5 Properties of $\succeq$

This section discusses the standard properties of $\succeq$. We do not take a stand on their intuitive plausibility, and just put them on the table.

Definition 25 (Properties of $\succeq$).
- *reflexivity:* $(\forall s)(s \succeq s)$
- *transitivity:* $(\forall s, t, u)\big((s \succeq t \wedge t \succeq u) \Rightarrow s \succeq u\big)$
- *totalness:* $(\forall s, t)(s \succeq t \vee t \succeq s)$
- *limitedness:* $\|\phi\| \neq \emptyset \Rightarrow \text{opt}_{\succeq}(\|\phi\|) \neq \emptyset$.

Totalness yields reflexivity as a special case.

Intuitively, limitedness and totalness play a different role: limitedness rules out infinite sequences of strictly better states ("no best"); totalness rules out incomparabilities in the ordering ("no gaps").

Remark 9. Limitedness has been popularized by Lewis under the name "limit assumption". It has been given other forms in the literature. These other forms will not concern us here.

Remark 10. Even though intuitively the two conditions of totalness and limitedness play a different role, on the formal side they are related to one another. First, even though limitedness does not imply totalness, a limited ordering may be transformed into a total one in a truth-preserving way (viz. without affecting the set of validities). Second, for limitedness to rule out infinite sequences of strictly better ϕ-states, $\succeq$ must be assumed to be total. These are more advanced topics, which fall outside the scope of this chapter.

Consider this alternative evaluation rule for the dyadic obligation operator:

$$s \vDash \bigcirc(\psi/\phi) \text{ iff } \neg\exists t \ (t \vDash \phi) \text{ or} \atop \exists t \ (t \vDash \phi \wedge \psi \ \& \ \forall u \ (u \succeq t \Rightarrow u \vDash \phi \rightarrow \psi)) \qquad (\exists\forall)$$

We shall refer to the statement appearing at the right-hand side of "iff" as the $\exists\forall$ rule. Intuitively, $\bigcirc(\psi/\phi)$ is true whenever either there is no ϕ-state, or there is an $\phi \wedge \psi$-state such that, as we go up in the ordering, the material implication $\phi \rightarrow \psi$ always holds. In other words, all states ranked as high as this state comply with the obligation in question.

Theorem 14. *The $\exists\forall$ rule implies the opt-rule. Given limitedness and transitivity, the opt-rule implies the $\exists\forall$ rule.*

Proof. Left as an exercise. $\square$

2.3.6   A simpler visual representation

Models become progressively more cumbersome to handle as the number of states increases. A simpler diagrammatic/visual representation in terms of "levels of ideality" (or, as a "layered" cake) is nevertheless possible if and only in so far as the betterness relation comes with the full panoply of the standard properties. The convention is that states within the same level are equally good, while being all strictly better than each state at a lower level. Ideality goes upwards: the higher a level is, the more ideal the states within it are.

The procedure for evaluating deontic formulas may be simplified thus. To check if $\bigcirc(\psi/\phi)$ holds, go the highest level containing states where the antecedent holds, and check if *all of these states* (where the antecedent holds) satisfy the consequent. If yes, $\bigcirc(\psi/\phi)$ holds. If no, $\bigcirc(\psi/\phi)$ does not hold. To check if $P(\psi/\phi)$ holds, go the highest level containing states where the antecedent holds, and check if *at least one of them* satisfy the consequent. If yes, $P(\psi/\phi)$ holds. If no, $P(\psi/\phi)$ does not hold.

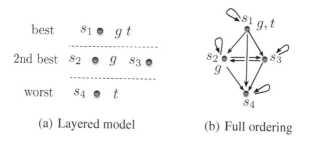

(a) Layered model   (b) Full ordering

Figure 2.3: Chisholm (a), (b)

This kind of diagrammatic representation was in fact used in section 2.1 to describe a typical preference model of the Chisholm set $\Gamma = \{\bigcirc g, \bigcirc(t/g), \bigcirc(\neg t/\neg g), \neg g\}$. Figure 2.3 puts side-by-side the layered model we gave and the "full" ordering it corresponds to. By a "full" ordering, we mean one that has been specified in full. Of course, if one is working with a reflexive and transitive betterness relation, the arrow representation can already be simplified to some extent, by omitting all the arrows that can be deduced by reflexivity and transitivity. Be that as it may, one can see that all of Γ is satisfied in, say, state s_3, and hence that Γ is consistent. Example 9 sustains this claim, by checking each obligation in Γ against the proposed model.

Example 9 (Consistency of the Chisholm set). Consider the picture to the left. $\bigcirc g$ holds, because g holds in the highest level. $\bigcirc(t/g)$ holds, because t holds in the highest level in which g holds. $\bigcirc(\neg t/\neg g)$ holds, because $\neg t$ holds in the highest level in which $\neg g$ holds. Consider the picture to the right. We have:

$$\text{opt}_{\succeq}(\|\top\| = W) = \{s_1\} \subseteq \|g\| \tag{2.1}$$
$$\text{opt}_{\succeq}(\|g\| = \{s_1, s_2\}) = \{s_1\} \subseteq \|t\| \tag{2.2}$$
$$\text{opt}_{\succeq}(\|\neg g\| = \{s_3, s_4\}) = \{s_3\} \subseteq \|\neg t\| \tag{2.3}$$

2.3.7 Examples of invalidities

In section 2.1, it is mentioned that $\bigcirc(\psi/\phi)$ aims at capturing a notion of conditional obligation that is "weaker" than the one obtained in MDL using $\rightarrow$ or $\dashv$3. "Weaker" means that the operator satisfies less laws than its MDL counterpart. In this section, we consider two well-known principles that fail for $\bigcirc(-/-)$: strengthening of the antecedent, and factual detachment.

Observation 1. *Strengthening of the Antecedent (SA) is invalid:*

$$\bigcirc(\psi/\phi) \rightarrow \bigcirc(\psi/\phi \wedge \chi) \tag{SA}$$

Proof. Let $M = (W, \preceq, V)$ be such that $W = \{s_1, s_2\}$, $\preceq$ is the reflexive closure of $\{(s_1, s_2)\}$, $V(p) = W$, $V(q) = \{s_1\}$, $V(r) = \{s_2\}$. (Recall that the reflexive closure of a binary relation R on a given set X is the smallest reflexive relation on X that contains R.) This is illustrated by figure 2.4.

Figure 2.4: Failure of SA

We have:

$$\text{opt}_{\succeq}(\|p\| = W) = \{s_1\} \subseteq \|q\| \Rightarrow s_i \models \bigcirc(q/p) \text{ for } i \in \{1, 2\}$$
$$\text{opt}_{\succeq}(\|p \wedge r\| = \{s_2\}) = \{s_2\} \nsubseteq \|q\| \Rightarrow s_i \nvDash \bigcirc(q/p \wedge r) \text{ for } i \in \{1, 2\}$$

$\square$

Observation 2. *Factual Detachment (FD) is not valid:*

$$(\bigcirc(\psi/\phi) \wedge \phi) \to \bigcirc\psi \tag{FD}$$

Proof. Let $M = (W, \preceq, V)$ be such that $W = \{s_1, s_2\}$, $\preceq$ is the reflexive closure of $\{(s_1, s_2)\}$, $V(p) = V(q) = \{s_2\}$. This is illustrated by figure 2.5.

Figure 2.5: Failure of FD

We have

$s_2 \models p$

$\text{opt}_{\succeq}(\|p\|) = \{s_2\} \subseteq \|q\| = \{s_2\} \Rightarrow s_2 \models \bigcirc(q/p)$

$\text{opt}_{\succeq}(\|\top\| = \{s_1\}) \not\subseteq \|q\| \Rightarrow s_2 \not\models \bigcirc q$

$\square$

It is not so much a matter of letting these two laws go than a matter of restricting their application. One can imagine cases where strengthening of the antecedent seems to be appropriate. A system not allowing any form of strengthening of the antecedent would be unable to account for these cases. The next section will introduce a restricted form of strengthening of the antecedent that is supported in DDL. It is called "Rational Monotony" in the non-monotonic logic literature [28]. Intuitively, it says that one can strengthen an antecedent when the added condition χ is permitted under the main condition, ϕ. Formally:

$$P(\xi/\phi) \wedge \bigcirc(\psi/\phi) \to \bigcirc(\psi/\phi \wedge \chi) \tag{RM}$$

Note, however, that the consequence relation associated with the logic remains monotonic. Similarly, a restricted form of factual detachment is supported in DDL, called "strong factual detachment" (SFD). This is the law:

$$(\bigcirc(\psi/\phi) \wedge \Box\phi) \rightarrow \bigcirc\psi \qquad \text{(SFD)}$$

A system not allowing any form of factual detachment fails to do its job, that of providing cues to action. Van Eck asks: "How can we take seriously a conditional obligation if it cannot, by way of detachment, lead to an unconditional obligation?" [52, p. 263]. Obligations are contextual and vary based on the setting. Consequently, an obligation always takes the form of a conditional statement. However, in the notation $\bigcirc(\psi/\phi)$, the antecedent ϕ has the nature of an hypothesis, which needs to be discharged for the obligation to apply, and lead to an action.

2.4   A hierarchy of systems

2.4.1   System E

Definition 26. **E** *is the proof system consisting of the following axiom schemas and rule schemas. (Labels are from [38].)*

ϕ, where ϕ is a tautology from PL	(PL)
$\Box(\phi \rightarrow \psi) \rightarrow (\Box\phi \rightarrow \Box\psi)$	($\Box$-K)
$\Box\phi \rightarrow \phi$	($\Box$-T)
$\neg\Box\phi \rightarrow \Box\neg\Box\phi$	($\Box$-5)
$\bigcirc(\psi \rightarrow \chi/\phi) \rightarrow (\bigcirc(\psi/\phi) \rightarrow \bigcirc(\chi/\phi))$	(COK)
$\bigcirc(\phi/\phi)$	(Id)
$\bigcirc(\chi/(\phi \wedge \psi)) \rightarrow \bigcirc((\psi \rightarrow \chi)/\phi)$	(Sh)
$\bigcirc(\psi/\phi) \rightarrow \Box \bigcirc (\psi/\phi)$	(Abs)
$\Box\psi \rightarrow \bigcirc(\psi/\phi)$	(Nec)
$\Box(\phi \leftrightarrow \psi) \rightarrow (\bigcirc(\chi/\phi) \leftrightarrow \bigcirc(\chi/\psi))$	(Ext)
If $\vdash \phi$ *and* $\vdash \phi \rightarrow \psi$ *then* $\vdash \psi$	(MP)
If $\vdash \phi$ *then* $\vdash \Box\phi$	($\Box$-Nec)

These axiom and rule schemas fall into three groups:

- First, we have axiom and rule schemas for $\Box$. These are $\Box$-K, $\Box$-T, $\Box$-5 and $\Box$-Nec. Taken together, they tell us that $\Box$ is a S5-modality;

- Second, we have axiom schemas for $\bigcirc(-/-)$: COK, Id and Sh. COK is the dyadic analogue of the K axiom. Id is the dyadic analogue of the principle of identity. Sh is named after Shoham, who first introduced it. Sh can be viewed as the deontic analogue of the deduction theorem;

- Third, we have axiom schemas governing the interplay between $\bigcirc(-/-)$ and $\Box$: Abs, Nec and Ext. Abs is Lewis's principle of absoluteness. It reflects the fact that the ranking is not made relative to states. Nec is the dyadic analogue of the principle of necessitation. Ext expresses a principle of replacement of logical equivalents.

Example 10. We have the following derived rule:

$$\text{If } \vdash_{\mathbf{E}} \phi \to \psi \text{ then } \vdash_{\mathbf{E}} \bigcirc(\phi/\xi) \to \bigcirc(\psi/\xi) \qquad (2.4)$$

Below: the required derivation (the subscript $\mathbf{E}$ is omitted):

1. $\vdash \phi \to \psi$	Assumption
2. $\vdash \Box(\phi \to \psi)$	$\Box$-Nec, 1
3. $\vdash \Box(\phi \to \psi) \to \bigcirc(\phi \to \psi/\xi)$	Nec
4. $\vdash \bigcirc(\phi \to \psi/\xi)$	MP, 2,3
5. $\vdash \bigcirc(\phi \to \psi/\xi) \to (\bigcirc(\phi/\xi) \to \bigcirc(\psi/\xi))$	COK
6. $\vdash \bigcirc(\phi/\xi) \to \bigcirc(\psi/\xi)$	MP, 4,5

Remark 11. The dyadic counterparts of $\bigcirc$-4 and $\bigcirc$-5 are derivable in system $\mathbf{E}$:

$$\bigcirc(\psi/\phi) \to \bigcirc(\bigcirc(\psi/\phi)/\xi) \qquad (\text{D-}\bigcirc 4)$$

$$P(\psi/\phi) \to \bigcirc(P(\psi/\phi)/\xi) \qquad (\text{D-}\bigcirc 5)$$

Theorem 15 (Soundness). *System $\mathbf{E}$ is (strongly) sound with respect to the class of all preference models, the class of those in which $\succeq$ is reflexive, and the class of those in which $\succeq$ is total.*

Proof. For soundness with respect to the class of all preference models, it suffices to verify that the axioms are valid in the class of all preference models, and that the rules preserve validity with respect to this class. We only show that Sh is valid. Assume $s \models \bigcirc(\chi/(\phi \wedge \psi))$. Let $t \in \text{opt}_{\succeq}(\|\phi\|)$ and $t \models \psi$. Suppose, to reach a contradiction, that $t \notin \text{opt}_{\succeq}(\|\phi \wedge \psi\|)$. We have $t \models \phi$, since $t \in \text{opt}_{\succeq}(\|\phi\|)$. Hence, $t \models \phi \wedge \psi$. Therefore, for $t \notin \text{opt}_{\succeq}(\|\phi \wedge \psi\|)$ to be the case, there must be some u such that $u \models \phi \wedge \psi$ and $t \not\succeq u$. But $u \models \phi$. This contradicts the assumption that $t \in \text{opt}_{\succeq}(\|\phi\|)$. Hence, $t \in \text{opt}_{\succeq}(\|\phi \wedge \psi\|)$. The opening assumption then yields, $t \models \chi$, which suffices for proving the claim, viz. $s \models \bigcirc(\chi/(\phi \wedge \psi))$.

For soundness with respect to the class of models in which $\succeq$ is reflexive, and the class of those in which $\succeq$ is total, it is enough to point out that none of the axioms and none of the rules call upon the properties of reflexivity or totalness. □

Theorem 16 gives an answer to a long-standing open problem.

Theorem 16 (Completeness). **E** *is (strongly) complete with respect to the class of all preference models, the class of those in which $\succeq$ is reflexive, and the class of those in which $\succeq$ is total.*

Proof. The proof of completeness with respect to the class of all preference models is complex, and requires a detour through an alternative modelling, flavoured by Chellas and Stalnaker, in terms of selection functions. For more details, see [39]. Completeness with respect to the class of preference models in which $\succeq$ is reflexive, and with respect to the class of those in which $\succeq$ is total, is established, by showing that, starting with a given preference model M with an arbitrary betterness relation $\succeq$, one can transform M into an "equivalent" model M' in which $\succeq$ is total, and hence reflexive. Equivalent means: the two models validate exactly the same formulas. For details, see [39]. □

Remark 12. Theorems 15 and 16 mean that reflexivity and totalness of $\succeq$ do not have any import, in the sense that they do not affect the logic.

Theorem 17 also gives an answer to a long-standing open problem.

Theorem 17 (Decidability). *The theoremhood problem in* **E** *is decidable.*

Proof. See [40]. ∎

2.4.2 System F

Definition 27. **F** *is the proof system obtained by supplementing* **E** *with*

$$\Diamond\phi \to (\bigcirc(\psi/\phi) \to P(\psi/\phi)) \qquad (\bigcirc\text{-D}^\star)$$

$\bigcirc$-D* is the dyadic analogue of the $\bigcirc$-D axiom. Intuitively, it rules out the possibility of conflicts between conditional obligations, for a "consistent" ϕ.

Remark 13. In **F**, the modal operator $\Box$ turns out to be "superfluous". This is because it becomes definable in terms of $\bigcirc(-/-)$. That is, $\Box\phi$ is equivalent to $\bigcirc(\bot/\neg\phi)$.

Remark 14. Recall that $\bigcirc\phi$ is short for $\bigcirc(\phi/\top)$. In **F**, $\bigcirc$ thus defined satisfies all the axioms of system **DS5** (cf. section 1.5).

Theorems 18 and 19 give an answer to two other long-standing open problems in deontic logic.

Theorem 18 (Soundness and completeness). **F** *is (strongly) sound and (strongly) complete with respect to the class of preference models in which $\succeq$ is limited, the class of those in which $\succeq$ is limited and reflexive, and the class of those in which $\succeq$ is limited and total.*

Proof. For soundness, it is enough to verify that, given the assumption of limitedness, $\bigcirc$-D* is valid. Let $M = (W, \succeq, V)$ be a preference model in which the betterness relation is required to be limited and consider some $s \in W$ such that $s \vDash \Diamond\phi$ and $s \vDash \bigcirc(\psi/\phi)$. From the former, there is $t \in W$ such that $t \vDash \phi$. From the latter, $\text{opt}_{\succeq}(\|\phi\|) \subseteq \|\psi\|$. Since $t \vDash \phi$, $\|\phi\| \neq \emptyset$. Hence, by the assumption of limitedness, $\text{opt}_{\succeq}(\|\phi\|) \neq \emptyset$. Since $\text{opt}_{\succeq}(\|\phi\|) \subseteq \|\psi\|$ and $\text{opt}_{\succeq}(\|\phi\|) \neq \emptyset$, $\text{opt}_{\succeq}(\|\phi\|) \cap \|\psi\| \neq \emptyset$, by set-theory. Hence,

$s \vDash P(\psi/\phi)$. This shows that $\bigcirc$-$\mathbf{D}^*$ is valid in the class of all preference models in which the betterness relation is required to be limited.

For soundness with respect to the class of models in which $\succeq$ is limited and reflexive, and the class of those in which $\succeq$ is limited and total, the argument is the same as in the proof for $\mathbf{E}$.

For completeness, see [39].   □

Theorem 19 (Decidability). *The theoremhood problem in* $\mathbf{F}$ *is decidable.*

Proof. See [40].   □

2.4.3   System G

Definition 28. $\mathbf{G}$ *is the proof system obtained by supplementing* $\mathbf{F}$ *with*

$$P(\psi/\phi) \wedge \bigcirc((\psi \to \chi)/\phi) \to \bigcirc(\chi/(\phi \wedge \psi)) \qquad \text{(Sp)}$$

Remark 15. Sp appears in Spohn's own axiomatization of Hansson's system DSDL3 (see [48]). Sp is equivalent to the so-called principle of rational monotony mentioned in section 2.3.7:

$$P(\psi/\phi) \wedge \bigcirc(\chi/\phi) \to \bigcirc(\chi/\phi \wedge \psi) \qquad \text{(RM)}$$

Example 11. We have:

$$\vdash_{\mathbf{G}} P(r/p) \wedge \bigcirc(q/p) \to \bigcirc(q/p \wedge r) \qquad (2.5)$$

Below: the required derivation:

1. $\vdash q \to (r \to q)$ PL
2. $\vdash \Box(q \to (r \to q))$ $\Box$-Nec, 1
3. $\vdash \Box(q \to (r \to q)) \to \bigcirc(q \to (r \to q)/p)$ Nec
4. $\vdash \bigcirc(q \to (r \to q)/p)$ MP, 2,3
5. $\vdash \bigcirc(q \to (r \to q)/p) \to (\bigcirc(q/p) \to \bigcirc(r \to q/p))$ COK
6. $\vdash \bigcirc(q/p) \to \bigcirc(r \to q/p)$ MP, 4,5
7. $\vdash \bigcirc(r \to q/p) \to (P(r/p) \to \bigcirc(q/p \wedge r))$ Sp, PL
8. $\vdash P(r/p) \wedge \bigcirc(q/p) \to \bigcirc(q/p \wedge r)$ PL, 6, 7

Theorem 20 (Soundness and completeness). **G** *is (strongly) sound and (strongly) complete with respect to the class of preference models in which* $\succeq$ *is limited and transitive, and with respect to the class of those in which it is limited, transitive and total (and hence reflexive).*

Proof. For soundness, it is enough to verify that, given transitivity, Sp is valid. Suppose not. There is then some $M = (W, \succeq, V)$, with $\succeq$ transitive, such that (i) $\text{opt}_{\succeq}(\|\phi\|) \subseteq \|\psi \to \chi\|$, (ii) $\text{opt}_{\succeq}(\|\phi\|) \cap \|\psi\| \neq \emptyset$, and (iii) $\text{opt}_{\succeq}(\|\phi \wedge \psi\|) \not\subseteq \|\chi\|$. From (iii), there is some t such that $t \in \text{opt}_{\succeq}(\|\phi \wedge \psi\|)$ and $t \not\models \chi$. From (i), $t \notin \text{opt}_{\succeq}(\|\phi\|)$, because $t \models \psi \wedge \neg\chi$. But $t \models \phi$. So there is some $u \models \phi$ with $t \not\succeq u$. From (ii), there is also some u' such that $u' \in \text{opt}_{\succeq}(\|\phi\|)$ and $u' \models \psi$. Since $u' \models \phi \wedge \psi$, $t \succeq u'$. Also, $u' \succeq u$, since $u' \in \text{opt}_{\succeq}(\|\phi\|)$. By transitivity, $t \succeq u$. Contradiction. Hence Sp is valid as long as $\succeq$ is transitive.

For completeness, see [38]. □

Remark 16. Theorem 20 tells us that, given limitedness and transitivity, the assumption of totalness of $\succeq$ has no import, in the sense that it does not affect the logic.

Theorem 21 (Decidability). *The theoremhood problem in* **G** *is decidable.*

Proof. See [40]. □

2.5 Notes

Hansson, most notably through [22], is broadly acknowledged as the father of DDL. The framework was developed further by a number of authors, including van Fraassen [53], Spohn [48], Lewis [29], Åqvist [4, 5], Hansen [18], Goble [16] and Parent [36, 38, 39, 40]. The history of the field is intricately intertwined with developments in three other related areas:

- rational choice theory: Sen [47].
- logic for counterfactuals: Lewis [29].
- non-monotonic logic: KLM systems [27, 28] .

For more information on the interplay between these areas, the reader is referred to Makinson [30].

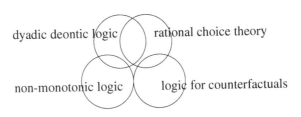

dyadic deontic logic   rational choice theory

non-monotonic logic   logic for counterfactuals

Figure 2.6: A bird's eye view

The opt-rule is due to Spohn. Hansson himself used a different–albeit related–evaluation rule. For details, see Parent [38, 39]. The ∃∀ rule is due to Lewis. He provided an axiomatization for the class of models equipped with a reflexive, transitive and total betterness relation. Goble [14] discovered an axiomatization for the case where the totalness requirement is relaxed.

The role of the deontic version of Rational Monotony is explained in relation to CTDs by Parent [40]. Prakken and Sergot[46] adopt Strong Factual Detachment as an intuitively valid principle for deontic conditionals.

Systems **E**, **F** and **G** were formulated by Åqvist. They echo Hansson's systems DSDL1, DSDL2 and DSDL3, respectively. The completeness and decidability of **G** has been known for some time. Spohn discovered a sound, weakly complete and decidable axiomatisation of Hansson's system DSDL3. The strong completeness and decidability of **E** and **F** have been established recently, in Parent [39].

2.6   Exercises

Exercise 14 (section 2.3, subsection 2.3.1). Give the satisfaction conditions for dyadic P (permission), monadic O and monadic P. That is, fill in the blank at the right-hand side of "iff":

- $M, s \vDash P(\psi/\phi)$ iff
- $M, s \vDash \bigcirc \phi$ iff

- $M, s \vDash P\phi$ iff

They should be analogous to the satisfaction conditions for dyadic $\bigcirc$ in definition 20, with *best* left undefined formally.

Exercise 15 (section 2.3, subsection 2.3.2). The aim of the exercise is construct a typical model of the set Γ of obligations described below. We assume this is done by counting violations. The example, discussed by Belzer [6] among others, concerns an instruction to officials accompanying Reagan and Gorbachov during so-called Reykjavík summit, on telling them a certain secret. p and q are for "telling Reagan" and "telling Gorbachov", respectively.

$$\Gamma = \{\bigcirc(\neg p \wedge \neg q), \bigcirc(p/q), \bigcirc(q/p)\}$$

For the purpose of the exercise, we assume that there are four states s_1-s_4, with p true at s_3 and s_4 and q true at s_2 and s_4. Answer these questions:

- How many obligations in Γ each state violate?
- What is the respective rank of the states?

Finally, explain why each obligation in Γ holds in this model ("holds" means "is true/satisfied in an arbitrarily chosen state").

Exercise 16 (section 2.3, subsection 2.3.5). Explain why totalness yields reflexivity. Explain why, given totalness, limitedness rules out infinite sequences of strictly better states.

Exercise 17 (section 2.3, subsections 2.3.4 and 2.3.5). Give the proof of theorem 14.

Exercise 18 (section 2.3, subsections 2.3.3 and 2.3.5). Show that $\succ$ is irreflexive. Show that, if $\succeq$ is transitive, then $\succ$ is transitive too. Show that, if $\succeq$ is assumed to be reflexive and transitive, then $\simeq$ is an equivalence relation, viz. $\simeq$ is reflexive, transitive and symmetric (if $s \simeq t$, then $t \simeq s$).

Exercise 19 (section 2.3, subsection 2.3.6). Consider figure 2.7. Write in full the ordering $\succeq$. Assume that *best* is spelled out in terms of the opt-rule. Check whether or not the following holds:

$$s_1 \vDash \bigcirc(r/p) \tag{2.6}$$

$$s_1 \models \bigcirc(q/p) \tag{2.7}$$
$$s_1 \models \bigcirc(r/p \wedge q) \tag{2.8}$$
$$s_1 \models \bigcirc(p \vee q/s) \tag{2.9}$$
$$s_1 \models P(r/p) \tag{2.10}$$
$$s_1 \models P(r/p \leftrightarrow q) \tag{2.11}$$

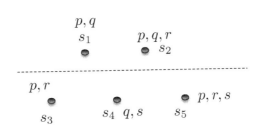

Figure 2.7: Levels of ideality

Exercise 20 (section 2.3, subsection 2.3.6, Chisholm). Determine if Γ meets the requirement of logical independence mentioned in the previous chapter.

Exercise 21 (section 2.4, subsections 2.4.1 and 2.4.2). Show that the axiom schema $\bigcirc$-D* is not valid in the class of preference models in which $\succeq$ is required to be reflexive. Explain why it follows from this that $\bigcirc$-D* is not a theorem of **E**.

Exercise 22 (section 2.4, subsection 2.4.1). Complete the proof of soundness of **E**.

Exercise 23 (section 2.4, subsection 2.4.1). Show that D-$\bigcirc$4 and D-$\bigcirc$5 are derivable in **E**. (For the second, you may find it helpful to use the following derived rule, which holds in S5: If $\vdash \phi \rightarrow \Box\psi$, then $\vdash \Diamond\phi \rightarrow \psi$.)

Exercise 24 (section 2.4, subsections 2.4.1 and 2.4.2). In **F**, the modal operator $\Box$ turns out to be "superfluous". This is because it becomes definable in terms of $\bigcirc(-/-)$. That is, $\Box\phi$ is equivalent to $\bigcirc(\bot/\neg\phi)$. This can be shown using the proof-theory, or using the semantics. This does not apply to the weaker system **E**. For this exercise, you are asked to demonstrate these claims using the semantics. To be more specific, you are asked to show that

- $\Box\phi \equiv \bigcirc(\bot/\neg\phi)$ is valid in the class of preference models in which the betterness relation is required to be limited;
- $\Box\phi \equiv \bigcirc(\bot/\neg\phi)$ is not valid in the class of all preference models.

Exercise 25 (section 2.4, subsections 2.4.2 and 2.4.3). Show that Sp is not valid in the class of preference models in which $\succeq$ is required to be reflexive and limited. Explain why it follows from this that Sp is not a theorem of **F**.

Part II

Norm-based Deontic Logic: I/O Logic

Chapter 3

Unconstrained I/O Logic

3.1 Introduction

This chapter and the next one deal with the input/output (I/O) logic developed by Makinson and van der Torre [32, 33, 34]. The basic idea is to generalize the theory of conditional norms from modal logic to the abstract study of conditional codes viewed as sets of relations between Boolean formulas.

The semantics is operational rather than truth-conditional. The meaning of the deontic concepts is given in terms of a set of procedures yielding outputs for inputs. The basic mechanism underpinning these procedures is that of detachment or modus-ponens. The associated proof-theory is formulated in terms of derivation rules operating on pairs of formulas rather than individual formulas.

This chapter focuses on the I/O analysis of the concept of obligation as described in Makinson and van der Torre [32]. This is the simplest case; no extra machinery is used to filter out the output. What is delivered is (as Makinson [31] calls it) the "gross output".

45

3.2　Preliminaries

Let $\mathbb{L}$ be the set of all the formulas of classical propositional logic. A normative system N is a set of (ordered) pairs (a, x) of formulas. Intuitively, a pair (a, x) denotes a conditional obligation. It is also called a rule. (a, x) is read as "given a, it ought to be the case that x". a is called the body (or antecedent), and x the head (or consequent). $(\top, x)$ denotes the unconditional obligation of x, where $\top$ is a tautology.

Given a set A of formulas (input set), we use the notation $out(N, A)$ to denote the output of A under N. Intuitively, the output of A under N is the set of obligations that apply to the current situation. We also use the equivalent notation $(A, x) \in out(N)$. When A is a singleton set, curly brackets are omitted.

⚠ Do not confuse $x \in out(N, a)$ with $(a, x) \in N$.

You may wonder how this new formalism relates with those studied in the previous chapters. The link with MDL can be made by defining the monadic obligation operator as follows:

$$N \models \bigcirc x \text{ iff } x \in out(N, \top)$$

The link with DDL can be made by defining the dyadic obligation operator as follows:

$$N \models \bigcirc(x/a) \text{ iff } x \in out(N, a)$$

We now introduce the two building blocks that will be used to define the I/O operations: the notion of consequence, and the notion of image.

Definition 29 (Consequence). *$Cn(A)$ denotes the set of logical consequences of A in classical propositional logic. That is, $Cn(A) = \{x : A \vdash x\}$ (for $\vdash$, read 'proves').*

Theorem 22. *Cn is a closure operation, viz it satisfies the properties:*

$$A \subseteq Cn(A) \qquad \qquad \text{(Inclusion)}$$
$$A \subseteq B \Rightarrow Cn(A) \subseteq Cn(B) \qquad \text{(Monotony)}$$
$$Cn(A) = CnCn(A) \qquad \qquad \text{(Idempotence)}$$

Remark 17. Cn satisfies the property of compactness:

$$x \in Cn(A) \Rightarrow \exists \text{ finite } A' \subseteq A \mid x \in Cn(A')$$

Definition 30 (Image). *The image of A under N (notation: $N(A)$) is the set $\{x : (a, x) \in N \text{ for some } a \in A\}$. For $N(A)$, read "the N of A".*

Intuitively: $N(A)$ gathers all the heads of the rules triggered by A in the sense that their body is an element of A. This is single step detachment.

Example 12.

N	A	$N(A)$
$\{(a_1, x_1), (a_2, x_2)\}$	$\{a_1\}$	$\{x_1\}$
$\{(a_1, x_1), (a_2, x_2)\}$	$\{a_1, x_2\}$	$\{x_1\}$
$\{(a_1, x_1), (a_2, x_2)\}$	$\{a_1, a_2\}$	$\{x_1, x_2\}$
$\{(a_1, x_1), (a_2, x_2)\}$	$\emptyset$	$\emptyset$

The bigger an input set is, the more heads can be detached from it:

Theorem 23 (Monotony). $A \subseteq B \Rightarrow N(A) \subseteq N(B)$.

Proof. Assume $A \subseteq B$. Let $x \in N(A)$. By definition of N, there is some $(a, x) \in N$ with $a \in A$. The opening assumption yields $a \in B$. By definition of N, it then follows that $x \in N(B)$. This shows that $N(A) \subseteq N(B)$. $\qquad\qquad\qquad\qquad\qquad\qquad\qquad\qquad\qquad\qquad\square$

3.3 Simple-minded I/O operation

3.3.1 Semantics

The operation out_1 (called "simple-minded") spells out the basic mechanism used to calculate the output in the non-iterated case. The general procedure is illustrated by figure 3.1.

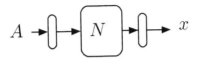

Figure 3.1: Basic mechanism

There are three different parts: there is the explicitly promulgated code, there is a pre-processor that prepares the input before it goes in, and a post-processor that unpacks it on the other side. In the case of out_1, one gets the following three steps. First, you close the input under Cn, viz. you determine the set of all logical consequences of the input. Next, you detach the heads of the rules whose bodies are in the latter set. Third, you close the set obtained at step two under Cn again.

Definition 31. *We define* $out_1(N, A)$ *as*

$$Cn(N(Cn(A)))$$

$$out_1(N) = \{(A, x) : x \in out_1(N, A)\}.$$

Example 13. Let $N = \{(a, x), (a \vee b, y)\}$. Put $A = \{a\}$.

A	$Cn(A)$	$N(Cn(A))$	out_1
a	$Cn(a)$	$\{x, y\}$	$Cn(x, y)$

Remark 18. $x \in out_1(N, A)$ says: $A \vdash \bigwedge_{i=0}^{n} a_i$ and $\bigwedge_{i=0}^{n} x_i \vdash x$, where $(a_1, x_1), ..., (a_n, x_n) \in N$.

3.3.2 Proof theory

A derivation of a pair (a, x) from N, given a set R of rules, is understood to be a tree with (a, x) at the root, each non-leaf node related to its immediate parents by the inverse of a rule in R, and each leaf node an element of N.

Definition 32. $(a, x) \in deriv_1(N)$ *if and only if* (a, x) *is derivable from* N *using the rules* $\{\top, SI, WO, AND\}$. *SI and WO abbreviate "strengthening of the input" and "weakening of the output", respectively.*

$$\top \; \frac{\overline{\quad\quad}}{(\top, \top)}$$

$$SI \; \frac{(a, x) \quad\quad b \vdash a}{(b, x)}$$

$$WO \; \frac{(a, x) \quad\quad x \vdash y}{(a, y)}$$

$$AND \; \frac{(a, x) \quad\quad (a, y)}{(a, x \wedge y)}$$

Given a set A of formulas, $(A, x) \in deriv_1(N)$ whenever $(a, x) \in deriv_1(N)$ for some conjunction a of formulas in A.

Put $deriv_1(N, A) = \{x : (A, x) \in deriv_1(N)\}$.

Example 14. Let $N = \{(a \vee b, x)\}$. We have $(b, x \vee y) \in deriv_1(N)$. The derived derivation can be displayed as a tree diagram, as follows:

$$WO \; \frac{SI \; \dfrac{(a \vee b, x)}{(b, x)}}{(b, x \vee y)}$$

3.3.3 Soundness and completeness

Theorem 24. out_1 *validates the rules of* $deriv_1$ *(for input* a*).*

⚠ That, e.g., SI is validated means that $x \in out_1(N, b)$ whenever both $x \in out_1(N, a)$ and $b \vdash a$.

Proof. We show SI. Assume $x \in out_1(N, a)$ and $b \vdash a$. From the first, $x \in Cn(N(Cn(a)))$, definition 31. From the second, $a \in Cn(b)$, definition of Cn. This is equivalent with $\{a\} \subseteq Cn(b)$. By monotony and idempotence for Cn, $Cn(a) \subseteq Cn(b)$. By monotony for N, one gets $N(Cn(a)) \subseteq N(Cn(b))$, and then $Cn(N(Cn(a))) \subseteq Cn(N(Cn(b)))$, by monotony for Cn again. Hence, $x \in Cn(N(Cn(b)))$. By definition 31, $x \in out_1(N, b)$. $\qquad\square$

Corollary 2 (Soundness). $deriv_1(N) \subseteq out_1(N)$.

Theorem 25 (Completeness). $out_1(N) \subseteq deriv_1(N)$.

Proof. See [32]. $\qquad\square$

3.4   Basic I/O operation

The operation out_2 (called "basic") injects (OR) into out_1 so that reasoning by cases is supported.

$$\text{OR } \frac{(a, x) \qquad (b, x)}{(a \vee b, x)}$$

3.4.1   Semantics

Throughout this section, V is a set of formulas. We say that V extends V' if $V' \subseteq V$, and that V is a proper extension of V' if $V' \subset V$.

Definition 33. V *is maximal consistent if*

$$V \nvdash \perp, and \tag{3.1}$$
$$y \notin V \Rightarrow V \cup \{y\} \vdash \perp \tag{3.2}$$

Intuitively: V is consistent, and no proper extension of V is consistent.

For some notation, MCS and MCE abbreviate "maximal consistent set" and "maximal consistent extension", respectively.

Fact 1. If V is a MCS, then

$$Cn(V) = V \tag{3.3}$$
$$\text{Either } b \in V \text{ or } \neg b \in V \tag{3.4}$$
$$\text{If } b \vee c \in V \text{ then: } b \in V \text{ or } c \in V \tag{3.5}$$

(3.3) says that V is closed under Cn. Facts (3.4) and (3.5) are called "$\neg$-completeness" and "saturatedness" (or "primeness"), respectively.

V is called complete if V is a MCS or equal to $\mathbb{L}$.

Definition 34. *We define* $out_2(N, A)$ *as*

$$\cap \{ Cn(N(V)) : A \subseteq V, V \text{ complete } \}$$

There is always at least one complete V *extending* A, *namely* $\mathbb{L}$ *itself. Remember also from propositional logic the so-called Lindenbaum's lemma: every consistent set is a subset of a maximal consistent set.*

As before, $out_2(N) = \{(A, x) : x \in out_2(N, A)\}$.

Example 15. Let $N = \{(a, x), (b, x)\}$ and $A = \{a \vee b\}$.

V	$N(V)$	$Cn(N(V))$
$\mathbb{L}$	$\{x\}$	$Cn(x)$
A given MCE of $\{a \vee b, a\}$	$\{x\}$	$Cn(x)$
A given MCE of $\{a \vee b, b\}$	$\{x\}$	$Cn(x)$

$\hookdownarrow out_2(N, A) = Cn(x) \cap Cn(x) \cap Cn(x) = Cn(x)$.

The difficulty lies in identifying the relevant complete sets, besides $\mathbb{L}$. In this example, we first consider a given MCE of $\{a \vee b, a\}$, and then a given MCE of $\{a \vee b, b\}$. (Both MCS's exist, by Lindenbaum's lemma). This choice is dictated by the fact that $a \vee b$ is given. Fact 1 (3.5) guarantees that a or b (or both) is in V.

Remark 19. Because $V = Cn(V)$, out_2 may be rephrased thus:

$$out_2(N, A) = \cap \{ out_1(N, V) : A \subseteq V, V \text{ complete} \}$$

3.4.2 Proof theory

Definition 35. $(a, x) \in deriv_2(N)$ *if and only if* (a, x) *is derivable from* N *using the rules of* $deriv_1$ *along with OR.*

As before $(A, x) \in deriv_2(N)$ if $(a, x) \in deriv_2(N)$ for some conjunction a of formulas in A. And $deriv_2(N, A) = \{x : (A, x) \in deriv_2(N)\}$.

3.4.3 Soundness and completeness

Theorem 26. out_2 *validates the rules of* $deriv_2$ *(for input a).*

Proof. We only show OR. Let $x \in out_2(N, a)$ and $x \in out_2(N, b)$. To show: $x \in out_2(N, a \vee b)$.

Let V be a complete set containing $a \vee b$. By definition of a complete set, i) either $a \in V$ or ii) $b \in V$. In case i), the assumption $x \in out_2(N, a)$ allows us to conclude $x \in Cn(N(V))$. In case ii), the assumption $x \in out_2(N, b)$ allows us to conclude $x \in Cn(N(V))$. Either way, $x \in Cn(N(V))$, which suffices for $x \in out_2(N, a \vee b)$. $\square$

Corollary 3 (Soundness). $deriv_2(N) \subseteq out_2(N)$.

Theorem 27 (Completeness). $out_2(N) \subseteq deriv_2(N)$.

Proof. See [32]. $\square$

3.5 Reusable I/O operation

The operation out_3 (called "reusable output") extends out_1 to iterations of successive detachments. Outputs may be recycled as inputs, and the rule CT ("Cumulative Transitivity") is validated:

$$\text{CT} \;\frac{(a, x) \qquad (a \wedge x, y)}{(a, y)}$$

Graphically:

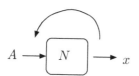

Figure 3.2: Recycling outputs as inputs

3.5.1 Semantics

Definition 36. *We define* $out_3(N, A)$ *as*

$$\cap\{Cn(N(B)) : A \subseteq B = Cn(B) \supseteq N(B)\}$$

There is always at least one such B, namely $\mathbb{L}$.

As before, we define $out_3(N) = \{(A, x) : x \in out_3(N, A)\}$.

Remark 20. Instead of considering in turn all the B's meeting the stated requirements, you may find it convenient to work with the smallest such B, call it $B^\star$–it is the meet of all the B's in question. Definition 36 can then be simplified into: $out_3(N, A) = Cn(N(B^\star))$, where $B^\star$ is the smallest B such that $A \subseteq B = Cn(B) \supseteq N(B)$. This alternative definition has a fixed-point flavour.

Example 16. Let $N = \{(a, x), (a \wedge x, y)\}$ and $A = \{a\}$. We have:

$B^\star$	$N(B^\star)$	$Cn(N(B^\star))$
$Cn(a, x, y)$	$\{x, y\}$	$Cn(x, y)$

$\hookrightarrow out_3(N, A) = Cn(x, y)$.

Below: an inductive characterization of out_3.

Definition 37 (Bulk increments, [49]). *Define* $out_3^b(N, A) = \cup_{i=0}^{\omega} A_i$ *where*

$$A_0 = out_1(N, A)$$
$$A_{i+1} = Cn(A_i \cup out_1(N, A_i \cup A))$$

Intuitively: output is continuously recycled as input, detaching heads of rules whenever possible.

All the A_i's are linearly ordered under $\subseteq$.

Example 17. In example 16, we have

$$A_0 = Cn(x)$$
$$A_1 = Cn(Cn(x) \cup out_1(N, Cn(x) \cup \{a\}))$$
$$\quad = Cn(Cn(x) \cup Cn(x, y))$$
$$\quad = Cn(x, y)$$
$$A_2 = A_1$$
$$\vdots$$

Theorem 28. out_3 *and* out_3^b *are equivalent.*

Proof. This is [49, Th. 4.3.12 and Th. 4.3.13].   □

3.5.2   Proof theory

Definition 38. $(a, x) \in deriv_3(N)$ *if and only if* (a, x) *is derivable from* N *using the rules of* $deriv_1$ *supplemented with CT.*

As before $(A, x) \in deriv_3(N)$ if $(a, x) \in deriv_3(N)$ for some conjunction a of formulas in A. And $deriv_3(N, A) = \{x : (A, x) \in deriv_3(N)\}$.

Remark 21. Plain transitivity is a derived rule:

$$\text{CT} \ \cfrac{(a, x) \quad \cfrac{(x, y)}{(a \wedge x, y)} \, \text{SI}}{(a, y)}$$

3.5.3   Soundness and completeness

Theorem 29. out_3 *validates the rules of* $deriv_3$ *(for input* a).

Proof. We only show CT. Let $x \in out_3(N, a)$ and $y \in out_3(N, a \wedge x)$. To show: $y \in out_3(N, a)$.

Let $B \mid a \in B = Cn(B) \supseteq N(B)$. The assumption $x \in out_3(N, a)$ yields $x \in Cn(N(B))$. But $N(B) \subseteq B = Cn(B)$. By monotony for Cn, $Cn(N(B)) \subseteq Cn(B)$, so that $x \in Cn(B)$, hence $x \in B$, from which one gets $a \wedge x \in B$. The assumption $y \in out_3(N, a \wedge x)$ now yields $y \in Cn(N(B))$, which suffices for $y \in out_3(N, a)$. $\quad\square$

Corollary 4 (Soundness). $deriv_3(N) \subseteq out_3(N)$.

Theorem 30 (Completeness). $out_3(N) \subseteq deriv_3(N)$.

Proof. See [32]. $\quad\square$

3.6 Basic reusable I/O operation

The operation out_4 (called "basic reusable") injects OR into out_3 so that reasoning by cases is also supported.

3.6.1 Semantics

Definition 39. *We define* $out_4(N, A)$ *as*

$$\cap\{Cn(N(V)) : A \subseteq V \supseteq N(V), V \text{ complete}\}$$

Define $out_4(N) = \{(A, x) : x \in out_4(N, A)\}$.

Remark 22. The definition of out_4 parallels that of out_2 but adds one extra requirement: $V \supseteq N(V)$.

Example 18. Let $N = \{(a, x), (a \wedge x, y)\}$ and $A = \{a\}$. We have:

V	$N(V)$	$Cn(N(V))$
$\mathbb{L}$	$\{x, y\}$	$Cn(x, y)$
A given MCE of $\{a, x, y\}$	$\{x, y\}$	$Cn(x, y)$

$\hookrightarrow out_4(N, A) = Cn(x, y)$.

Example 19. Let $N = \{(a, x), (b, x)\}$ and $A = \{a \vee b\}$.

V	$N(V)$	$Cn(N(V))$
$\mathbb{L}$	$\{x\}$	$Cn(x)$
A given MCE of $\{a \vee b, x\}$	$\{x\}$	$Cn(x)$

↳ $out_4(N, A) = Cn(x)$.

Remark 23. $out_4(N, A) = out_2(N, A \cup m(N))$ where $m(N)$ is the set of all materialisations of elements of N, i.e., the set of all formulas $b \rightarrow y$ with $(b, y) \in N$.

3.6.2   Proof theory

Definition 40. $(a, x) \in deriv_4(N)$ *if and only if* (a, x) *is derivable using the rules of* $deriv_2$ *supplemented with CT.*

As before $(A, x) \in deriv_4(N)$ if $(a, x) \in deriv_4(N)$ for some conjunction a of formulas in A. And $deriv_4(N, A) = \{x : (A, x) \in deriv_4(N)\}$.

3.6.3   Soundness and completeness

Theorem 31. out_4 *validates the rules of* $deriv_4$ *(for input a).*

Corollary 5 (Soundness). $deriv_4(N) \subseteq out_4(N)$.

Theorem 32 (Completeness). $out_4(N) \subseteq deriv_4(N)$.

Proof. See [32].   □

3.7   Concluding remarks

Table 3.1 shows the I/O operations and the associated rules. For each of these operations, we may also consider a throughput version that also allows inputs to reappear as outputs. These are the operations $out_i^+(N, A) = out_i(N^+, A)$, where $N^+ = N \cup I$ and I is the set of all pairs (a, a) for formulas a. It turns out that $out_4^+(N, A) = Cn(A \cup m(N))$, thus collapsing into classical logic.

Output operation	Rules
Simple-minded (out_1)	$\{\top, \text{SI}, \text{WO}, \text{AND}\}$
Basic (out_2)	$\{\top, \text{SI}, \text{WO}, \text{AND}\}+\{\text{OR}\}$
Reusable (out_3)	$\{\top, \text{SI}, \text{WO}, \text{AND}\}+\{\text{CT}\}$
Basic reusable (out_4)	$\{\top, \text{SI}, \text{WO}, \text{AND}\}+\{\text{OR},\text{CT}\}$

Table 3.1: I/O systems

3.8 Notes

The norm-based approach to deontic logic goes back to Makinson [31], and has roots in Alchourrón and Bulygin [1]'s analysis of normative systems. Here are two other examples of a norm-based deontic logic: Hansen [19, 20]'s imperativist logic and Horty [26]'s theory of reasons.

Variant I/O systems are possible. Two of them deserve special mention:

- Intuitionistic I/O logics [41]. These are defined on top of intuitionistic logic instead of classical logic.
- Aggregative I/O logics [43]. These do not contain WO, and use a form of transitivity other than CT, which may be called "aggregative cumulative transitivity" (ACT). This is the rule:

$$\text{ACT} \ \frac{(a, x) \qquad (a \wedge x, y)}{(a, x \wedge y)}$$

The I/O logic approach to permission is described in Makinson and van der Torre [34]. A semantics based on formal concept analysis is given in Stolpe [50]. An overview of I/O logics is given in Parent and van der Torre [42].

3.9 Exercises

Exercise 26 (section 3.2). Explain the difference between $x \in out(N, a)$ and $(a, x) \in N$.

Exercise 27 (section 3.2). Express the set of all the tautologies in the Cn notation.

Exercise 28 (section 3.2). Let $N = \{(a, x), (x, y), (x \wedge y, z)\}$. What is $N(N(\{a, x, y\}))$?

Exercise 29 (section 3.2). What is $N(\mathbb{L})$?

Exercise 30 (section 3.3). Let $N = \{(a, x), (b, y), (a \vee b, z)\}$. What is $out_1(N, \{a, b\})$? Let $N = \{(\top, x)\}$. What is $out_1(N, a)$?

Exercise 31 (section 3.3). Complete the proof of soundness of out_1. That is, show that out_1 validates AND, WO and $\top$.

Exercise 32 (section 3.3). Show that ID ("identity") and CONT ("contraposition") fail for out_1:

$$\text{ID } \frac{-}{(a, a)} \qquad\qquad \text{CONT } \frac{(a, x)}{(\neg x, \neg a)}$$

Exercise 33 (section 3.3). For $N = \{(\top, x), (a, y \wedge z)\}$, do we have $(a, x \wedge z) \in deriv_1(N)$?

Exercise 34 (section 3.4). Show that out_2 validates the rules of $deriv_1$.

Exercise 35 (section 3.4). Show that OR fails for out_1.

Exercise 36 (section 3.4). Put $N = \{(a \wedge x, y), (a \wedge \neg x, y)\}$. What is $out_2(N, a)$?

Exercise 37 (section 3.4). Suppose $N = \{(a, x), (b, y)\}$. Show that $(a \vee b, x \vee y) \in deriv_2(N)$.

Exercise 38 (section 3.5). Show that out_3 validates the rules of $deriv_1$ (in addition to CT).

Exercise 39 (section 3.5). Show that OR fails for out_3.

Exercise 40 (section 3.6). Show that out_4 validates the rules of $deriv_4$.

Exercise 41 (section 3.6). Show that out_4 validates Ghost Contraposition (GC):

$$\text{GC} \frac{(\neg x, \neg a) \qquad (a \wedge x, y)}{(a, y)}$$

Exercise 42 (sections 3.5 and 3.8). Explain informally the difference between ATC and CT. Show that, in the presence of AND and WO, the two rules are equivalent.

Chapter 4

Filtering Excess Output

4.1 Constrained I/O logic

Constrained I/O logic ("cIOL", for short) aims at giving a finer grained analysis of the notion of obligation than unconstrained I/O logic does. This is achieved by using constraints to filter out excess output. Here we proceed in two stages, and introduce a second I/O operation on top of the initial one. Because of the lack of a corresponding proof-theory for the second I/O operation, in section 4.2 these two stages will be merged into one.

The following two problems have led to the use of constraints in input/output logics:

- the question of how to deal with violations and obligations resulting from violations, known as contrary-to-duty (CTD) reasoning. It has been discussed in the context of the notorious contrary-to-duty paradoxes such as Chisholm's paradox and Forrester [11]'s paradox;

- the question of how to accommodate deontic dilemmas (unsolvable conflicts between obligations), and the question of how to reason about conflicting obligations of different strengths.

The use of constraints has been introduced in [33] in relation to CTDs, and has been extended to the topic of conflicts in [37].

4.1.1   Norm violation

The following two examples show why contrary-to-duty scenarios create a problem in I/O logic without constraints.

Example 20 (Chisholm). Let $out = out_i$, where $i \in \{3, 4\}$. Assume $N = \{(\top, h), (h, t), (\neg h, \neg t)\}$, where h and t are for *helping* and *telling*. Put $A = \{\neg h\}$. We have $out(N, A) = Cn(t, \neg t) = \mathbb{L}$.

Example 21 (Forrester [11]). Let $out = out_i$, where $i \in \{1, 2, 3, 4\}$. Assume $N = \{(\top, \neg k), (k, k \wedge g)\}$, where k and $k \wedge g$ are for *killing* and *killing-gently*. Put $A = \{k\}$. We have $out(N, A) = \mathbb{L}$.

The strategy is to adapt a technique that is well known in the logic of belief change—cut back the set of norms to just below the threshold of yielding excess, and consider the resulting output. This amounts to carrying out a contraction on the set N of norms.

The above strategy is implemented using a set C of formulas (called constraints) as an extra parameter. It is used to filter out excessive output. What we get is (as Makinson calls it) the "net" output (as opposed to the gross output).

Definition 41 (Maxfamily). *Let $out = out_i$, where $i \in \{1, 2, 3, 4\}$. We define*

- *maxfamily(N, A, C) is the set of $\subseteq$-maximal subsets N' of N such that $out(N', A)$ is consistent with C;*
- *outfamily$(N, A, C) =$*
  $\{out(N', A) \mid N' \in maxfamily(N, A, C)\}$.

For CTDs, it is assumed that $C = A$ (I/O constraint). The input represents something that is unalterably true, viz. something that has been done and cannot be changed anymore (cf. Hansson [22, §13]'s interpretation of circumstances). The output must remain consistent with it.

In general a set of norms and an input do not have a set of formulas as output, but a set of sets of formulas. We can infer a set of formulas by taking the join (credulous) or the meet (skeptical).

Definition 42 (Constrained, net output). *Define*

$$out_c(N, A) = out_{\cup/\cap}(N, A) = \cup/\cap \, outfamily(N, A, C)$$

⚠ In most CTD scenarios, the choice between $\cup$ and $\cap$ does not arise, because the maxfamily has one element only.

Example 22 (Chisholm, cont'd). Let *out*, N and A be as in example 22. Put $C = A$. We have $maxfamily(N, A, C) = \{\{(h, t), (\neg h, \neg t)\}\}$, and so $out_c(N, A) = Cn(\neg t)$.

By contrast, $maxfamily(N, \{h\}, \{h\}) = \{\{(\top, h), (h, t), (\neg h, \neg t)\}\}$, and so $out_c(N, A) = Cn(h, t)$.

⚠ One can describe what is going on in example 22, by saying that, given input $\neg h$, the primary obligation $(\top, h)$ is first removed from N (in belief revision theory, this is called a contraction), because it generates an inconsistency.

Example 23 (Forrester, cont'd). Let *out*, N and A be as in example 21. Put $C = A$. We have $maxfamily(N, A, C) = \{\{(k, k \wedge g)\}\}$, and so $out_c(N, A) = Cn(k \wedge g)$.

By contrast, $maxfamily(N, \{\neg k\}, \{\neg k\}) = \{\{(\top, \neg k), (k, k \wedge g)\}\}$, and so $out_c(N, A) = Cn(\neg k)$.

A notable feature of the proposed approach is that, in a violation context, the primary obligation "disappears". We will get back to it in section 4.2.

4.1.2 Accommodating dilemmas

Example 24 illustrates why dilemmas (unsolvable conflicts between obligations) create a problem in I/O logic without constraints.

Example 24 (Unary conflict). Let $out = out_i$, where $i \in \{1, 2, 3, 4\}$. Assume $N = \{(a, b), (a, \neg b)\}$ and $A = \{a\}$. We have $out(N, A) = Cn(b, \neg b) = \mathbb{L}$.

To accommodate dilemmas, we use the same strategy as for CTDs. For simplicity's sake, we assume that the final output is obtained by taking the meet (skeptical).

Example 25 (Unary conflict, cont'd). Let *out*, N and $A = \{a\}$ be as in example 24. Put $C = \emptyset$. The maxfamily has two elements, $\{(a, b)\}$ and $\{(a, \neg b)\}$. The outfamily has two elements, $Cn(b)$ and $Cn(\neg b)$. So $out_c(N, A) = Cn(\emptyset)$.

Example 26 (Binary conflict). Let $out = out_i$, where $i \in \{1, 2, 3, 4\}$. Assume $N = \{(a, b), (a, c)\}$, $A = \{a\}$ and $C = \{b \to \neg c\}$. We have $out_c(N, A) = Cn(b \vee c)$.

There are three main desiderata for a logic admitting the possibility of normative conflicts. They are written below in the modal logic notation.

Desideratum 1 Make conflicts logically consistent

$$\bigcirc a, \bigcirc \neg a \nvdash \bot$$

Desideratum 2 Avoid deontic explosion

$$\bigcirc a, \bigcirc \neg a \nvdash \bigcirc b$$

Desideratum 3 Do not give away too much, that is account for the validity of any other inference pattern that may seem unobjectionable at face value, like

$$\bigcirc (a \vee b), \bigcirc \neg a \vdash \bigcirc b$$

The I/O approach meets these desiderata. This is illustrated below with the example of desideratum 3.

Example 27 (Alternative service, Horty [26]). Let *out* be an I/O operation. Assume $N = \{(\top, f \vee s), (\top, \neg f)\}$, where f and s are for *fighting* *in the army* and *performing an alternative national service*, respectively. Put $A = \{\top\}$ and $C = \emptyset$. We have $out_c(N, A) = Cn(s, \neg f)$. Hence

$$f \vee s \in out_c(N, A)$$
$$\neg f \in out_c(N, A)$$
$$s \in out_c(N, A)$$

Link with Poole systems

There are connections between cIOL and some well-known systems for nonmonotonic reasoning developed in AI. In particular a so-called Poole system (and, as Makinson calls it, its associated "default assumption consequence" relation) may be seen as a special case of cIOL.

We recall the basic idea underpinning a Poole system: when an inconsistency is generated one looks at what follows from all the maximally consistent (more briefly, maxiconsistent) subsets of the set K of background assumptions.

Let (K, A, C) be a Poole system. K, A and C are sets of formulas in the language of classical propositional logic. K is a set of background assumptions, A is the input, and C is a set of constraints.

Definition 43 (Extfamily). *Given a Poole system (K, A, C), the family of its extensions in the sense of Poole is written as extfamily(K, A, C). Formally:*

- *extfamily(N, A, C) is the set of $Cn(A \cup K')$, where K' is maximal (w.r.t. $\subseteq$) among the subsets K'' of K such that $A \cup K'' \cup C$ is consistent.*

Theorem 33. *Given some (K, A, C), define $N = \{(\top, x) : x \in K\}$, and let outfamily be defined using the reusable basic throughput I/O operation out_4^+. We have that*

$$extfamily(K, A, C) = outfamily(N, A, C)$$

Proof. This is [33, Observation 4]. Hint for the proof: $out_4^+(N, A)$ equates $Cn(A \cup m(N))$. $\qquad \square$

4.1.3   Obligations of different strengths

This section defines a procedure for resolving conflicts between obligations based on a relation of priority between them. We start with the basic idea underpinning the procedure.

> **Basic idea** Start with a priority relation $\geq$ on pairs in N. Lift it to a priority relation $\geq^s$ on sets of pairs. Use $\geq^s$ to select a "preferred" element in the maxfamily. Restrict the final, net output to this preferred element.

Let $\geq \subseteq N \times N$. $(a, x) \geq (b, y)$ is read: (a, x) is at least as strong as (b, y). $\geq$ is required to be reflexive and transitive. (a, x) and (b, y) are said to be incomparable under $\geq$ if $(a, x) \not\geq (b, y)$ and $(b, y) \not\geq (a, x)$.

$>$ is the strict order induced by $\geq$. $(a, x) > (b, y)$ is read: (a, x) is strictly stronger than (b, y). $>$ is defined by putting $(a, x) > (b, y)$ whenever $(a, x) \geq (b, y)$ and $(b, y) \not\geq (a, x)$.

Lifting

Different notions of lifting have been considered in the literature. For the purposes of this book we shall not take a stand on whether one definition is ultimately the "correct" one. We just contrast two of them. The first one is named after Brass.

Definition 44 (Brass lifting). *$N_1 \geq^s_b N_2$ if and only if*

$$\forall (a, x) \in N_2 - N_1 \; \exists (b, y) \in N_1 - N_2 \text{ such that } (b, y) \geq (a, x)$$

Definition 45 ($\forall\forall$ lifting). *$N_1 \geq^s_{\forall\forall} N_2$ if and only if*

$$\forall (a, x) \in N_1 \forall (b, y) \in N_2 \; (a, x) \geq (b, y)$$

The definition of $>^s_b$ (resp. $>^s_{\forall\forall}$) in terms of $\geq^s_b$ (resp. $\geq^s_{\forall\forall}$) is analogous to the definition of $>$ in terms of $\geq$:
- $N_1 >^s_b N_2$ if and only if $N_1 \geq^s_b N_2$ but $N_2 \not\geq^s_b N_1$
- $N_1 >^s_{\forall\forall} N_2$ if and only if $N_1 \geq^s_{\forall\forall} N_2$ but $N_2 \not\geq^s_{\forall\forall} N_1$

Here is an example where the two definitions give a different result.

Example 28 (Mission, adapted from Goble [15]). Let a, b and c be three actions (for instance, three missions to perform). Put:

$$N = \{(\top, a), (\top, b), (\top, c)\} \tag{4.1}$$

Assume $(\top, a) > (\top, b) > (\top, c)$. Put $C = \{a \rightarrow (b \rightarrow \neg c), a \rightarrow \neg b\}$ and $A = \{\top\}$. Intuitively, only a and c can be done simultaneously, or b and c. The maxfamily has two elements, $N_1 = \{(\top, a), (\top, c)\}$ and $N_2 = \{(\top, b), (\top, c)\}$. N_1 and N_2 are not comparable under $\geq^s_{\forall\forall}$, while the Brass definition yields that $N_1 >^s_b N_2$. The second outcome seems more intuitive. Intuitively, the best option combined with the worst one outranks the second-best option combined with the worst one.

In the next section, we use $\geq^s_b$ and its strict counterpart $>^s_b$.

Preferred output

Definition 46 (Preferred output, out_P). *Let preffamily(N, A, C) be the set of $\geq^s_b$-maximal elements of maxfamily(N, A, C). Formally, this is the set*

$$\{N' \in maxfamily(N, A, C) : \forall N'' \in maxfamily(N, A, C),$$
$$N'' \geq^s_b N' \Rightarrow N' \geq^s_b N''\}$$

The preferred output is defined as follows:

$$out_P(N, A) = \cap/\cup \{out(N, A) : N \in preffamily(N, A, C)\}$$

Note that in typical examples there is only one element in the pref-family.

Not-triggered high-ranking obligations raise special issues as illustrated below. Examples 29 and 30 describe a case where the least important obligation overrules a more important one. This, in order to avoid the violation of an even more important obligation.

Example 29 (Cancer, [37]). Assume $out = out_i$, where $i \in \{3, 4\}$. Let $(b, c) > (a, b) > (a, \neg b)$, where a, b and c denote a set of data, the fact

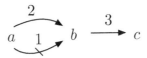

Figure 4.1: Cancer

of having chemo, and the fact of keeping the white blood cell (WBC) count high enough using a drug. This is illustrated by figure 4.1. An arrow of the form $x \longrightarrow y$ indicates that, according to the normative system, y is obligatory given x. A positive rule $(x, \neg y)$, with a negated proposition as its head, is written as the negative link $x \rightsquigarrow y$. Numbers show priorities.

Put $A = C = \{a\}$. The maxfamily has two elements, $N_1 = \{(b, c), (a, \neg b)\}$ and $N_2 = \{(b, c), (a, b)\}$. The preffamily has one element, N_2. And so $out_P(N, A) = Cn(b, c)$.

Example 30 (Cancer, cont'd, [37]). Let the norms and priorities involved be as in Example 29. But put $A = C = \{a, \neg c\}$. Intuitively, we are in a case where the physician knows that the drug to be used to prevent too much loss of white blood cells will remain ineffective. The maxfamily has two elements, $N_1 = \{(b, c), (a, \neg b)\}$ and $N_2 = \{(a, b)\}$. The preffamily has one element, N_1. And so $out_P(N, A) = Cn(\neg b)$. This tallies with our intuitions. A low white blood cell count is one of the more serious negative effects of chemotherapy. When the WBC count cannot be kept at a sufficient level, usually chemotherapy is stopped temporarily. $(a, \neg b)$ is lower in rank than (a, b). Still, the former overrides the latter. This is because the fulfilment of (a, b) triggers (b, c), which is even higher in rank. By going for b, the agent would put himself in a violation state with respect to an even more important norm.

The order puzzle has been introduced by Horty in [25].

Example 31 (Order puzzle, [25]). Let $(a, \neg b) > (\top, b) > (\top, a)$, where a and b are for *putting the heating on* and *opening the window* respectively. The norms $(\top, a)$, $(\top, b)$ and $(a, \neg b)$ are issued by a priest, a bishop and a cardinal, respectively. This is illustrated by figure 4.2.

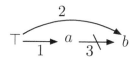

Figure 4.2: Order puzzle

Assume $out = out_i$, where $i \in \{3, 4\}$. Put $A = \{\top\}$ and $C = \emptyset$. The maxfamily has three elements, $N_1 = \{(\top, a), (\top, b)\}$ and $N_2 = \{(\top, a), (a, \neg b)\}$ and $N_3 = \{(\top, b), (a, \neg b)\}$. We have $N_3 >_b^s N_2 >_b^s N_1$. So the preffamily has one element, N_3. And so $out_P(N, A) = Cn(b)$.

Intuitively, this may be justified thus. There are three candidate outputs:

$$Cn(a, b) \quad Cn(a, \neg b) \quad Cn(b)$$

$Cn(a, b)$ is disregarded, because it leads to a violation of $(a, \neg b)$, which has the highest rank. $Cn(a, \neg b)$ is also disregarded, because $(\top, b)$ has a higher priority than $(\top, a)$, and the former overrides the latter in context $\top$. Furthermore, it is only if a is the case that $(a, \neg b)$ is triggered.

4.2 I/O logic with a consistency check

A drawback of constrained I/O logic is that (like Reiter's default logic) it has no known proof theory. This is due to the use of (external) constraints to filter excess output modulo a consistency check. In this section, we show how to get a surrogate of proof theory, by making the consistency check part of the semantic definition of the I/O operation. To be more precise, we define a variant of the simple-minded I/O operation out_1 and of the reusable I/O operation out_3 with a built-in consistency check. The I/O operations are written O_1 and O_3, respectively. The built-in consistency check gets reflected in the proof theory in the form of a consistency proviso restraining the application of a rule.

4.2.1   Single-step detachment

We start with the simplest case where outputs are not recycled as inputs.

Let $x \dashv\vdash y$ be a shorthand for $x \vdash y$ and $y \vdash x$. Given a set M of pairs, let $b(M)$ be the set of all bodies of elements of M, and $h(M)$ be the set of all heads of elements of M. Given some input set A, we say that a pair of the form (b, y) in M is "directly" grounded in A whenever $A \vdash b$.

Intuitively, definition 47 may be glossed as follows. Given input A, x is outputted if the following conditions hold: there is a non-empty and finite $M \subseteq N$, whose elements are all directly grounded in A; x is logically equivalent to the conjunction of all the heads of all the pairs in M; all the bodies of all the pairs in M are "collectively" consistent with x. M may be be called the "witness" for x. It gathers all the pairs effectively used to get x.

Formally:

Definition 47 (Semantics). $x \in O_1(N, A)$ *iff there is some finite* $M \subseteq N$ *and a set* $B \subseteq Cn(A)$ *such that* $M \neq \emptyset$, $B = b(M)$, $x \dashv\vdash \wedge h(M)$ *and* $\{x\} \cup B$ *is consistent.*

As before $O_1(N) = \{(A, x) : x \in O_1(N, A)\}$.

Example 32. Put $N = \{(\top, a), (\neg a, \neg a)\}$. We have $a \in O_1(N, \{\neg a\})$. Also, $\neg a \in O_1(N, \{\neg a\})$. The following table tells us why.

	input A	witness M	B	potential output x	consistency check
1.	$\{\neg a\}$	$\{(\top, a)\}$	$\{\top\}$	a	pass
2.	$\{\neg a\}$	$\{(\neg a, \neg a)\}$	$\{\neg a\}$	$\neg a$	pass

The top row deals with output a. The witness is $\{(\top, a)\}$. This pair is directly grounded in the input, and its body is consistent with the output. The bottom row deals with output $\neg a$. The witness is $\{(\neg a, \neg a)\}$. This pair is directly grounded in the input, and its body is consistent with the output.

On the other hand, $a \wedge \neg a \notin O_1(N, A)$. Witness:

	input A	witness M	B	potential output x	consistency check
3.	$\{\neg a\}$	N	$\{\top, \neg a\}$	$a \wedge \neg a$	fail

There are three possible choices of a witness M; only the one that puts $M = N$ is considered, because one wants the conjunction of the heads of the two pairs in N. For this choice of M, $B = \{\top, \neg a\}$. Our potential output is a contradiction, and thus the consistency check fails.

The corresponding proof theory has two salient features: WO goes away; a restricted version of AND holds.

Definition 48 (Proof system). $(a, x) \in D_1(N)$ *if and only if* (a, x) *is derivable from N using the rules* $\{SI, OEQ, R\text{-}AND\}$.

$$OEQ \; \frac{(a, x) \quad x \dashv\vdash y}{(a, y)}$$

$$R\text{-}AND \; \frac{(a, x) \quad (a, y) \quad a \wedge x \wedge y \text{ is consistent}}{(a, x \wedge y)}$$

Furthermore, for each leaf (b, y) of the derivation, $b \wedge y$ is required to be consistent.

OEQ stands for "output equivalence". R-AND stands for "restricted AND". In the terminology of Makinson and van der Torre [33], a pair (a, x) is said to have a consistent fulfilment if $a \wedge x$ is consistent. The rule R-AND says: the heads of two pairs with the same body may be joined by the conjunction *and* if and only in so far as the resulting pair (in the conclusion) has a consistent fulfilment.

As before, when A is a set of formulas, $(A, x) \in D_1(N)$ means that $(a, x) \in D_1(N)$, for some conjunction a of elements of A. Furthermore, $D_1(N, A) = \{x : (A, x) \in D_1(N)\}$.

Theorem 34 (Soundness and completeness). $O_1(N, A) = D_1(N, A)$

Proof. See [44]. □

4.2.2   Recycling outputs as inputs

We extend the I/O operation described in the previous section so that outputs can be recycled as inputs.

Compared with the previous treatment, the main difference is that a pair in witness M may be "indirectly" grounded in A (that is, modulo its chaining with other pairs in M that are themselves directly grounded in A). This is the import of clause i) in definition 49 below. One still checks for the consistency of x with the set of all the bodies of all the pairs in M that are directly grounded in A.

Definition 49 (Semantics). $x \in O_3(N, A)$ *iff there is a finite* $M \subseteq N$ *and a set* $B \subseteq Cn(A)$ *such that* $M(B) \neq \emptyset$, $x \dashv\vdash \wedge h(M)$, *and*

i) $\forall B' \big(B \subseteq B' = Cn(B') \supseteq M(B') \Rightarrow b(M) \subseteq B' \big)$
ii) $\{x\} \cup B$ *is consistent*

Example 33 (Möbius strip, [31]). Consider three conditional obligations stating that $\neg a$ is obligatory given c, that c is obligatory given b, and that b is obligatory given a, together with the fact that a is true:

$$N = \{(a, b), (b, c), (c, \neg a)\} \qquad A = \{a\}$$

For instance, a, b, c could say that Alice (respectively Bob, Carol) is invited to dinner. The obligation (a, b) says that if Alice is invited then Bob too should be invited, and so on. This is illustrated with figure 4.3.

Given input $A = \{a\}$, $b \wedge c$ is outputted, but not $b \wedge c \wedge \neg a$. The following table shows why. Intuitively, M still acts as the witness for output x, viz. the set of pairs used to get x (modulo chaining). B gathers the bodies of all the pairs in M that are directly grounded in the input A. The last but one column shows the meet of all the B's satisfying the equation $B \subseteq B' = Cn(B') \supseteq M(B')$.

	input A	witness M	B	potential output x	$\cap B'$	consistency check
1.	$\{a\}$	$\{(a, b), (b, c)\}$	$\{a\}$	$b \wedge c$	$Cn(a, b, c)$	pass
2.	$\{a\}$	N	$\{a\}$	$b \wedge c \wedge \neg a$	$\mathbb{L}$	fail

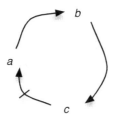

Figure 4.3: Möbius strip

Definition 50 (Proof system). $(a, x) \in D_3(N)$ *if and only if* (a, x) *is derivable from* N *using the rules* $\{SI, OEQ, R\text{-}ACT\}$:

$$R\text{-}ACT \frac{(a, x) \qquad (a \wedge x, y) \qquad a \wedge x \wedge y \text{ is consistent}}{(a, x \wedge y)}$$

As before it is required that all the leaves of the derivation of (a, x) *have a consistent "fulfilment". That is, for all the leaves* (b, y), $b \wedge y$ *must be consistent.*

R-ACT stands for "restricted aggregative cumulative transitivity". As before, when A is a set of formulas, $(A, x) \in D_3(N)$ means that $(a, x) \in D_3(N)$ for some conjunction a of elements of A. Furthermore, $D_3(N, A) = \{x : (A, x) \in D_3(N)\}$.

Remark 24. R-AND is derivable.

Theorem 35 (Soundness and completeness). $O_3(N, A) = D_3(N, A)$.

Proof. See [44]. ☐

4.2.3 A problem of Broersen and van der Torre

We illustrate the role of the consistency check further, with reference to a problem pointed out by Broersen and van der Torre [8]. They start by making the following observation:

"The pragmatic oddity is the derivation of the conjunction 'you should keep your promise and apologize for not keeping it' from 'you should keep your promise', 'if you do not keep your promise you should apologize' and 'you do not keep your promise' [45]. Note that the sentences of this problem have the same structure as those of the Chisholm scenario. The drowning problem is that many solutions of the pragmatic oddity cancel the obligation in case of violation, such that for violations $\neg p \wedge \bigcirc p$, the violated obligation $\bigcirc p$ is no longer derivable." [8, p. 64]

They go on to ask the following question:

"How to prevent the pragmatic oddity without creating the drowning problem?"

Example 34 shows that the proposed account provides an answer to their question. For the sake of simplicity, we adopt a proof-theoretical perspective, and focus on the simpler system D_1.

Example 34 (Broken promise). Let k and a stand for *keeping one's promise* and for *apologizing*, respectively. Consider the following derivation, in which a blocked derivation step is represented by a dashed line.

$$\text{SI} \, \frac{(\top, k)}{\underset{(\neg k, k \wedge a)}{\overline{(\neg k, k)} \quad \underline{(\neg k, a)}}} \, \text{R-AND}$$

On the one hand, the drowning problem does not occur, because SI allows us to move from $(\top, k)$ to $(\neg k, k)$. (Constrained I/O logic blocks such a move: k is not consistent with $\neg k$.) On the other hand, the pragmatic oddity is avoided, because R-AND cannot be applied to get $(\neg k, k \wedge a)$. This is because this pair does not have a consistent fulfilment.

4.3   Notes

The question of how to accommodate the possibility of conflicts between obligations is one of the benchmark problems of deontic logic.
The reader will find in the chapter by L. Goble [15] an excellent overview
of all the issues raised by this problem, with pointers to literature. The
three main desiderata mentioned in section 4.1.2 were identified by him.

The procedure for resolving conflicts described in section 4.1.3 owes
much to Hansen [19]'s logic of prioritized conditional imperatives and
to Horty [26]'s logic of reasons based on Reiter's default logic.

The use of constraints has another natural domain of application: the
analysis of permission viewed as an exception to an existing prohibition.
This topic is studied further by Boella and van der Torre [7]. An I/O
analysis of the notion of legal interpretation is given by Maranhao [35].

4.4   Exercises

Exercise 43 (section 4.1, subsection 4.1.1). Suppose $N = \{(a, b \wedge \neg x), (a \wedge b, x)\}$. What is $maxfamily(N, a, \emptyset)$, when a) $out = out_i$ where $i \in \{1, 2\}$, and when b) $out = out_i$ where $i \in \{3, 4\}$?

Exercise 44 (section 4.1, subsection 4.1.1). Let k, a and s stand for *keeping a promise, apologizing* and *feeling ashamed*, respectively. Put $N = \{(\top, k), (\neg k, a), (\neg k \wedge \neg a, s)\}$. What is $out_c(N, \top)$, $out_c(N, \neg k)$ and $out_c(N, \neg k \wedge \neg a)$?

Exercise 45 (section 4.1, subsection 4.1.2). Show that $b \in out_\cap(N, \top)$ whenever $b \vee c \in out_\cap(N, \top)$ and $\neg c \in out_\cap(N, \top)$.

Exercise 46 (section 4.1, subsection 4.1.3). What are the properties of $>$?

Exercise 47 (section 4.1, subsection 4.1.3). It is claimed that in example 28 i) N_1 and N_2 are not comparable under $\geq_{\vee\vee}^s$ and ii) $N_1 >_b^s N_2$. Justify these two claims.

Exercise 48 (section 4.1, subsection 4.1.3). State in full the definition of $>_b^s$.

Exercise 49 (section 4.2, subsections 4.2.1 and 4.2.2). Definitions 47 and 49 both require the leaves of a derivation to have a consistent fulfilment. Using an example, explain why this is needed.

Exercise 50 (section 4.2, subsection 4.2.1). Show that the rules WO and $\top$ fails for O_1. Show that SI, OEQ and R-AND hold for O_1.

Exercise 51 (section 4.2, subsection 4.2.3). Example 34 uses the proof theory to explain how Broersen and van der Torre's problem can be handled using a consistency check. Make the same point, but using the semantics, and more precisely O_3.

Exercise 52 (section 4.2). Reformulate in the I/O idiom the three desiderata for a logic accommodating conflicts mentioned in subsection 4.1.2. Show that both O_1 and O_3 meet these desiderata (although the third one is a variant form).

Exercise 53 (section 4.2, van Fraassen's paradox). The question of how to avoid deontic explosion is sometimes described as a matter of choosing between AND and WO. Prima facie the following derivation might serve the cause of defending this view.

$$\text{AND} \, \frac{\dfrac{(a,b) \qquad (a,\neg b)}{(a, b \wedge \neg b)}}{(a, c)} \, \text{WO}$$

With AND replaced by R-AND, the derivation is blocked. At the same time, desideratum 3 in subsection 4.1.2 is met. To let WO go as well may seem like an overkill, but it is not. This is because {R-AND, WO} still gives rise to deontic explosion. Show that there is a derivation of (a, c) from $\{(a, b), (a, \neg b)\}$ using these two rules. (This is called "van Fraassen's paradox" by van der Torre and Tan [51].)

Appendix A

Solutions to Selected Exercises

A.1 Chapter 1

Exercise 1.

1. $\bigcirc p$.
2. $\bigcirc (p \vee q)$. The alternative rendering $(\bigcirc p) \vee (\bigcirc q)$ would not do. The sentence may be true, if one disjunct only is true. Linguists make a distinction between "must" and "have to", but it cannot be capture with our rudimentary language.
3. $\bigcirc (\bigcirc \phi \to \phi)$. The use of a schematic letter captures the "whatever". ϕ could be an atomic proposition or a complex formula.

Exercise 3.

1. No: every state has a good successor.
2. We have $s_1 \models \bigcirc Pp$. This is because s_1 has two good successors, s_2 and s_3, each of which satisfies Pp. (s_2 satisfies Pp, because s_4 satisfies p and s_4 is a good successor of s_2. s_3 satisfies Pp, because it satisfies p and it is a good successor of itself.) It is not the case that $s_2 \models \bigcirc Pp$. This is because s_4 is a good successor of s_2, and it is not the case that $s_4 \models Pp$. (And it is not the case that $s_4 \models Pp$, because s_2 is the only good successor of s_4, and it does not satisfy p).
3. The states that satisfy the formula $\bigcirc p \to p$ are those that either

falsify $\bigcirc p$ (s_1, s_2 and s_4) or satisfy p (s_3).

Exercise 5.
- Direct proof. Let M be a model, and s a state in M, be such that $s \models \bigcirc\phi \vee \bigcirc\psi$. This means that either $s \models \bigcirc\phi$ or $s \models \bigcirc\psi$. We break the argument into cases.

  Case 1: $s \models \bigcirc\phi$. Let t be such that sRt. Since $s \models \bigcirc\phi$, $t \models \phi$, and so $t \models \phi \vee \psi$. Hence $s \models \bigcirc(\phi \vee \psi)$.

  Case 2: $s \models \bigcirc\psi$. A similar argument yields $s \models \bigcirc(\phi \vee \psi)$.

  Either way, $s \models \bigcirc(\phi \vee \psi)$. Hence, $\bigcirc\phi \vee \bigcirc\psi \models \bigcirc(\phi \vee \psi)$.
- Indirect proof (proof by contradiction). Let M be a model, and s a state in M, be such that $s \models \bigcirc\phi \vee \bigcirc\psi$, and $s \not\models \bigcirc(\phi \vee \psi)$. The latter means that there is t such that sRt and $t \not\models \phi \vee \psi$. Thus, $t \not\models \phi$ and $t \not\models \psi$. By the satisfaction condition for $\bigcirc$, $s \not\models \bigcirc\phi$ and $s \not\models \bigcirc\psi$, and so $s \not\models \bigcirc\phi \vee \bigcirc\psi$, a contradiction.

Exercise 7.
- Axiom schema ($\bigcirc$-D). Let M be a model, and s a state in M, be such that $s \models \bigcirc\phi$. This means that, for all t, if sRt, then $t \models \phi$. By the assumption of seriality, there is at least one such t. Hence, there is some t such that sRt and $t \models \phi$. By the satisfaction condition for P, $s \models P\phi$. In other words, $s \models \bigcirc\phi \rightarrow P\phi$. Since M and s and arbitrary, the axiom schema is valid.
- Rule schema ($\bigcirc$-Nec). Assume $\models \phi$. Let M be a model, and s and t two states in M such that sRt. Since $\models \phi$, $t \models \phi$. By the satisfaction condition for $\bigcirc$, $s \models \bigcirc\phi$. Since s and M are arbitrary, $\models \bigcirc\phi$, and hence the rule schema preserves validity.

Exercise 9.

1. $\vdash \phi \rightarrow (\bigcirc\phi \rightarrow \phi)$	PL
2. $\vdash \bigcirc\phi \rightarrow \bigcirc(\bigcirc\phi \rightarrow \phi)$	$\bigcirc$-K, $\bigcirc$-Nec, 1
3. $\vdash \neg\bigcirc\phi \rightarrow \bigcirc\neg\bigcirc\phi$	$\bigcirc$-5
4. $\vdash \neg\bigcirc\phi \rightarrow (\bigcirc\phi \rightarrow \phi)$	PL
5. $\vdash \bigcirc\neg\bigcirc\phi \rightarrow \bigcirc(\bigcirc\phi \rightarrow \phi)$	$\bigcirc$-K, $\bigcirc$-Nec, 3
6. $\vdash \neg\bigcirc\phi \rightarrow \bigcirc(\bigcirc\phi \rightarrow \phi)$	PL, 3,5
7. $\vdash \bigcirc(\bigcirc\phi \rightarrow \phi)$	PL, 2,6

Exercise 11.
- For □-Nec

1. $\vdash \phi$	Assumption
2. $\vdash \phi \to (\neg\phi \to v)$	PL
3. $\vdash \neg\phi \to v$	MP, 1,2
4. $\vdash \Box(\neg\phi \to v)$	□-Nec, 3

- For □-K: compact solution, using the derived rule:

$$\frac{\vdash \phi \wedge \phi' \to \psi}{\vdash \Box\phi \wedge \Box\phi' \to \Box\psi}\ RR$$

1. $\vdash (\neg(\phi \to \psi) \to v) \to ((\neg\phi \to v) \to (\neg\psi \to v))$	PL
2. $\vdash ((\neg(\phi \to \psi) \to v) \wedge (\neg\phi \to v)) \to (\neg\psi \to v))$	PL, 1
3. $\vdash \Box(\neg(\phi \to \psi) \to v) \wedge \Box(\neg\phi \to v) \to \Box(\neg\psi \to v))$	RR, 3
4. $\vdash \Box(\neg(\phi \to \psi) \to v) \to (\Box(\neg\phi \to v) \to \Box(\neg\psi \to v))$	PL, 4

Below: a derivation of RR:

1. $\vdash (\phi \wedge \phi') \to \psi$	Assumption
2. $\vdash \phi \to (\phi' \to \psi)$	PL, 1
3. $\vdash \Box\phi \to \Box(\phi' \to \psi)$	□-Nec, □-K, 2
4. $\vdash \Box\phi \to (\Box\phi' \to \Box\psi)$	□-K, 3
5. $\vdash (\Box\phi \wedge \Box\phi') \to \Box\psi$	PL, 4

Exercise 13.
1. Assume $\models \phi \to \psi$. Let M and s in M be such that $M, s \models \bigcirc\phi$. This means that $\|\phi\| \in N(s)$. From the opening assumption, $\|\phi\| \subseteq \|\psi\|$. By closure under superset, $\|\psi\| \in N(s)$, which suffices for $M, s \models \bigcirc\psi$. Hence $\models \bigcirc\phi \to \bigcirc\psi$.
2. $s \models P\phi$ iff $W - \|\phi\| \notin N(s)$; and $s \models F\phi$ iff $W - \|\phi\| \in N(s)$.

A.2 Chapter 2

Exercise 15.
- s_1: no violation.
- s_2 violates 2 obligations: $\bigcirc(\neg p \wedge \neg q)$; $\bigcirc(p/q)$.
- s_3 violates 2 obligations: $\bigcirc(\neg p \wedge \neg q)$; $\bigcirc(q/p)$.

- s_4 violates 1 obligation: $\bigcirc(\neg p \wedge \neg q)$.

Hence their ranks are: s_1 is best, s_4 is 2nd best, and s_2 and s_3 are worst. We have:

- $\|p\| = \{s_3, s_4\}$, $\|q\| = \{s_2, s_4\}$ and $\|\neg p \wedge \neg q\| = \{s_1\}$
- $best(\|p\|) = \{s_4\}$, $best(\|q\|) = \{s_4\}$ and $best(\|\top\|) = \{s_1\}$

Hence

- $best(\|\top\|) \subseteq \|\neg p \wedge \neg q\|$
- $best(\|p\|) \subseteq \|q\|$
- $best(\|q\|) \subseteq \|p\|$.

So, for all s_i $(1 \leq i \leq 4)$,

- $s_i \vDash \bigcirc(\neg p \wedge \neg q)$, $s_i \vDash \bigcirc(q/p)$ and $s_i \vDash \bigcirc(p/q)$.

Exercise 17.

- $\exists\forall$ rule $\Rightarrow$ opt-rule. Suppose $\neg\exists t$ $(t \vDash \phi)$ or $\exists t$ $(t \vDash \phi \wedge \psi$ & $\forall u$ $(u \succeq t \Rightarrow u \vDash \phi \rightarrow \psi))$. In the first case, $opt_\succeq(\|\phi\|) = \emptyset$, and so $opt_\succeq(\|\phi\|) \subseteq \|\psi\|$. In the second case, let $s' \in opt_\succeq(\|\phi\|)$. We have $s' \succeq t$ and $s' \vDash \phi$. So $s' \vDash \psi$, which suffices for $opt_\succeq(\|\phi\|) \subseteq \|\psi\|$ as required.
- Given limitedness and transitivity, opt-rule $\Rightarrow \exists\forall$ rule. Suppose $opt_\succeq(\|\phi\|) \subseteq \|\psi\|$. Either i) $opt_\succeq(\|\phi\|) = \emptyset$ or ii) $opt_\succeq(\|\phi\|) \neq \emptyset$. In case i), by limitedness, $\|\phi\| = \emptyset$, and so the $\exists\forall$ rule is verified. In case ii), there is some t such that $t \in opt_\succeq(\|\phi\|)$. We have $t \vDash \psi$, by the opening assumption. So $t \vDash \phi \wedge \psi$. Let u be such that $u \succeq t$ and $u \vDash \phi$. Consider any u' such that $u' \vDash \phi$. We have $t \succeq u'$. By transitivity, we then get $u \succeq u'$, so that $u \in opt_\succeq(\|\phi\|)$, and hence $u \vDash \psi$, by the opening assumption. Thus, the $\exists\forall$ rule is verified too.

Exercise 19. $\succeq$ is the reflexive and transitive closure of

$$\{(s_1, s_2), (s_2, s_1), (s_1, s_3),$$
$$(s_3, s_4), (s_4, s_3), (s_4, s_5), (s_5, s_4)\}$$

Remember that the transitive closure of a binary relation R on a set X is the smallest relation on X that contains R and is transitive. The following applies:

$$s_1 \nvDash \bigcirc(r/p) \text{ (witness: } s_1)$$

$$s_1 \models \bigcirc(q/p)$$
$$s_1 \not\models \bigcirc(r/p \wedge q) \text{ (witness: } s_1)$$
$$s_1 \models \bigcirc(p \vee q/s)$$
$$s_1 \models P(r/p)$$
$$s_1 \not\models P(r/p \leftrightarrow q) \text{ (witness: } s_1)$$

Exercise 21. Let $M = (W, \succeq, V)$ where $W = \{s_0, s_1\}$, $\succeq$ is the reflexive closure of W, $V(p) = W$ and $V(q) = \emptyset$. We have

- $s_0 \models \Diamond p$ since, e.g., $s_0 \models p$
- $s_0 \models \bigcirc(q/p)$ since $\text{opt}_\succeq(\|p\|) = \emptyset \subseteq \|q\| = \emptyset$
- $s_0 \not\models P(q/p)$ since $\text{opt}_\succeq(\|p\|) \cap \|q\| = \emptyset$.

So, $s_0 \not\models \Diamond p \rightarrow (\bigcirc(q/p) \rightarrow P(q/p))$. Hence, $\bigcirc$-D* is not valid in the class of all preference models in which $\succeq$ is required to be reflexive.

The soundness theorem for **E** tells us that, if a formula is a theorem of **E**, then it is valid in the class of preference models in which $\succeq$ is required to be reflexive. The contrapositive of the latter statement says: if a formula is not valid in the class of preference models in which $\succeq$ is required to be reflexive, then it is not a theorem of **E**. Applying this to $\bigcirc$-D*, one immediately gets that $\bigcirc$-D* is not a theorem of **E**.

Exercise 23.

- For D-$\bigcirc$4

1. $\vdash \bigcirc(\psi/\phi) \rightarrow \Box \bigcirc (\psi/\phi)$	Abs
2. $\vdash \Box \bigcirc (\psi/\phi) \rightarrow \bigcirc(\bigcirc(\psi/\phi)/\xi)$	Nec
3. $\vdash \bigcirc(\psi/\phi) \rightarrow \bigcirc(\bigcirc(\psi/\phi)/\xi)$	PL, 1, 2

- For D-$\bigcirc$5

1. $\vdash \bigcirc(\neg\psi/\phi) \rightarrow \Box \bigcirc (\neg\psi/\phi)$	Abs
2. $\vdash \Diamond \bigcirc (\neg\psi/\phi) \rightarrow \bigcirc(\neg\psi/\phi)$	inference rule of S5, 1
3. $\vdash \neg \bigcirc (\neg\psi/\phi) \rightarrow \neg\Diamond \bigcirc (\neg\psi/\phi)$	PL, 4
4. $\vdash P(\psi/\phi) \rightarrow \Box P(\psi/\phi)$	def of $\Diamond$ and P, 5
5. $\vdash \Box P(\psi/\phi) \rightarrow \bigcirc(P(\psi/\phi)/\xi)$	Nec
6. $\vdash P(\psi/\phi) \rightarrow \bigcirc(P(\psi/\phi)/\xi)$	PL, 6, 7

Exercise 25. Let $M = (W, \succeq, V)$ where

- $W = \{s_0, s_1, s_2\}$;
- $\succeq$ is the reflexive closure of
  $$\{(s_0, s_1), (s_1, s_0), (s_1, s_2), (s_2, s_1), (s_0, s_2)\};$$
- $V(p) = W$, $V(q) = \{s_0, s_1\}$ and $V(r) = \{s_1, s_2\}$.

This is illustrated by figure A.1, where reflexivity is omitted.

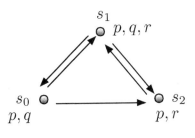

Figure A.1: Countermodel to Sp

$\succeq$ is reflexive, and it is not transitive since $(s_2, s_0) \notin \succeq$. To verify that $\succeq$ is limited, one must check that each non-empty subset X of W has an optimal element:

$X \subseteq W, X \neq \emptyset$	$\mathrm{opt}_\succeq(X)$
W	$\{s_0, s_1\}$
$\{s_0, s_1\}$	$\{s_0, s_1\}$
$\{s_0, s_2\}$	$\{s_0\}$
$\{s_1, s_2\}$	$\{s_1, s_2\}$
$\{s_0\}$	$\{s_0\}$
$\{s_1\}$	$\{s_1\}$
$\{s_2\}$	$\{s_2\}$

We have:

$$\mathrm{opt}_\succeq(\|p\|) = \{s_0, s_1\} \subseteq \|r \to q\|$$
$$\hookrightarrow s_i \models \bigcirc(r \to q/p) \text{ for } i \in \{0, 1, 2\}$$
$$\mathrm{opt}_\succeq(\|p\|) \cap \|q\| \neq \emptyset$$
$$\hookrightarrow s_i \models P(q/p) \text{ for } i \in \{0, 1, 2\}$$

$$\text{opt}_{\succeq}(\|p \wedge r\| = \{s_1, s_2\}) = \{s_1, s_2\} \not\subseteq \|q\| = \{s_0, s_1\}$$
$$\hookrightarrow s_i \not\models \bigcirc(q/p \wedge r) \text{ for } i \in \{0, 1, 2\}$$

This shows that Sp is not valid in the class of preference models in which $\succeq$ is required to be reflexive and limited. It follows that Sp is not a theorem of **F** for the same reason as in the solution to exercise 21. The soundness theorem for **F** tells us that, if Sp was a theorem of **F**, then Sp would be valid in the class of preference models in which $\succeq$ is required to be reflexive and limited. Since it is not valid in this class of models, it is not a theorem of **F**.

A.3 Chapter 3

Exercise 27. $Cn(\emptyset)$.

Exercise 29. $N(\mathbb{L})$ is the set of all the heads of all the pairs in N.

Exercise 31. We only show AND. Assume $x \in out_1(N, a)$ and $y \in out_1(N, a)$. By definition of out_1, one gets $x, y \in Cn(N(Cn(a)))$. By classical logic, one then gets $x \wedge y \in CnCn(N(Cn(a)))$. By idempotence, one gets $x \wedge y \in Cn(N(Cn(a))) = out_1(N, a)$.

Exercise 33. Yes.

$$\text{AND } \dfrac{\text{SI } \dfrac{(\top, x)}{(a, x)} \quad \text{WO } \dfrac{(a, y \wedge z)}{(a, y)}}{(a, x \wedge y)}$$

Exercise 35. Put $N = \{(a, x), (b, x)\}$. We both have $x \in out_1(N, a)$ and $x \in out_1(N, b)$, but $x \notin out_1(N, a \vee b)$. The reason for the latter is that neither $a \vee b \vdash a$ nor $a \vee b \vdash b$. In other words, no pair in N is triggered by $a \vee b$.

Exercise 37.

$$\text{WO } \dfrac{\text{OR } \dfrac{(a, x) \quad (b, x)}{(a \vee b, x)}}{(a \vee b, x \vee y)}$$

Exercise 39. Put $N = \{(a, x), (b, x)\}$. We have:

- $x \in out_3(N, a)$. Indeed:

$$\frac{B \qquad N(B) \qquad Cn(N(B))}{Cn(a, x) \qquad \{x\} \qquad Cn(x)}$$

- $x \in out_3(N, b)$. Indeed:

$$\frac{B \qquad N(B) \qquad Cn(N(B))}{Cn(b, x) \qquad \{x\} \qquad Cn(x)}$$

- $x \notin out_3(N, a \vee b)$. Indeed:

$$\frac{B \qquad N(B) \qquad Cn(N(B))}{Cn(a \vee b) \qquad \emptyset \qquad Cn(\emptyset)}$$

Exercise 41. Let i) $\neg a \in out_4(N, \neg x)$ and ii) $y \in out_4(N, a \wedge x)$. Let V be a complete set such that $a \in V \supseteq N(V)$. Assume V is maximal consistent. Note that $V \supseteq N(V)$ implies $Cn(V) = V \supseteq Cn(N(V))$. So if $\neg x \in V$, then i) would yield $\neg a \in Cn(N(V))$, from which $\neg a \in V$ would follow. Hence $\neg x \notin V$, and so $x \in V$. Thus $a \wedge x \in V$. By ii), $y \in Cn(N(V))$. Assume V is $\mathbb{L}$. In this case, $a \wedge x \in V$, and by ii) again $y \in Cn(N(V))$.

This shows that $y \in out_4(N, a)$.

A.4 Chapter 4

Exercise 43. For out_1 and out_2, the maxfamily is $\{N\}$. For out_3 and out_4, the maxfamily is $\{\{(a, b \wedge \neg x)\}, \{(a \wedge b, x)\}\}$.

Exercise 45. Assume $b \vee c \in out_\cap(N, \top)$ and $\neg c \in out_\cap(N, \top)$. Consider some $N' \in maxfamily(N, \top, C)$. By definition of maxfamily, we have $b \vee c \in out(N', \top)$ and $\neg c \in out(N', \top)$, where $out = out_i$ for $i \in \{1, 2, 3, 4\}$. Since out is closed under classical consequence, $b \in out(N', \top)$, which suffices for $b \in out_\cap(N, \top)$.

Exercise 47. On the one hand, $N_1 \not\geq^s_{\vee\vee} N_2$ as $(\top, c) \not\geq (\top, b)$. On the other hand, $N_2 \not\geq^s_{\vee\vee} N_1$ because $(\top, c) \not\geq (\top, a)$. This shows that N_1 and N_2 are incomparable under $\geq^s_{\vee\vee}$.

We have $N_2 - N_1 = \{(\top, b)\}$ and $N_1 - N_2 = \{(\top, a)\}$. Since $(\top, a) \geq (\top, b)$, $N_1 \geq_b^s N_2$. Since $(\top, b) \not\geq (\top, a)$, $N_2 \not\geq_b^s N_1$. Hence $N_1 >_b^s N_2$.

Exercise 49. Let $N = \{(a, \neg a)\}$. Without such a requirement, we would have $(a, \neg a) \in D_1(N)$ and $(a, \neg a) \in D_3(N)$. But $\neg a \notin O_1(N, a)$:

input A	witness M	B	potential output x	consistency check
$\{a\}$	N	$\{a\}$	$\neg a$	fail

Similarly, $\neg a \notin O_3(N, a)$ (the requirement $M(B) \neq \emptyset$ in the definition of O_3 dictates the choice of B):

input A	witness M	B	potential output x	B'	consistency check
$\{a\}$	N	$\{a\}$	$\neg a$	$\mathbb{L}$	fail

Hence without such a requirement the claims of soundness, theorems 34 and 35, would fail.

Exercise 51. Let $N = \{(\top, k), (\neg k, a)\}$. We have that $k \in O_3(N, \neg k)$. Also $a \in O_3(N, \neg k)$. But $k \wedge a \notin O_3(N, \neg k)$. Witness:

input A	witness M	B	potential output x	$\cap B'$	consistency check
$\{\neg k\}$	$\{(\top, k)\}$	$\{\top\}$	k	$Cn(k)$	pass
$\{\neg k\}$	$\{(\neg k, a)\}$	$\{\neg k\}$	a	$Cn(\neg k, a)$	pass
$\{\neg k\}$	$\{(\top, k), (\neg k, a)\}$	$-$	$k \wedge a$	$\mathbb{L}$	fail

Exercise 53.

$$\text{R-AND} \cfrac{\text{WO} \cfrac{(a, b)}{(a, b \vee c)} \qquad (a, \neg b)}{\cfrac{(a, (b \vee c) \wedge \neg b)}{(a, c)} \text{WO}}$$

Bibliography

[1] C. Alchourrón and E. Bulygin. *Normative Systems.* Springer-Verlag, New York and Vienna, 1971.

[2] A. Anderson. A reduction of deontic logic to alethic modal logic. *Mind*, 67(265):100–103, 1958.

[3] A. Anderson. Some nasty problems in the formal logic of ethics. *Noûs*, 1(4):345–360, 1967.

[4] L. Åqvist. *An Introduction to Deontic logic and the Theory of Normative Systems.* Bibliopolis, Naples, 1987.

[5] L. Åqvist. Deontic logic. In D. Gabbay and F. Guenthner, editors, *Handbook of Philosophical Logic*, volume 8, pages 147–264. Kluwer Academic Publishers, Dordrecht, Holland, 2nd edition, 2002.

[6] M. Belzer. Legal reasoning in 3-D. In *Proceedings of the 1st International Conference on Artificial Intelligence and Law*, ICAIL'87, pages 155–163, New York, NY, USA, 1987. ACM.

[7] G. Boella and L. van der Torre. Institutions with a hierarchy of authorities in distributed dynamic environments. *Artificial Intelligence and Law*, 16(1):53–71, 2008.

[8] J. Broersen and L. van der Torre. Ten problems of deontic logic and normative reasoning in computer science. In N. Bezhanishvili

and V. Goranko, editors, *Lectures on Logic and Computation*, volume 7388 of *Lecture Notes in Computer Science*, pages 55–88. Springer, Berlin, Heidelberg, 2012.

[9] B. Chellas. *Modal Logic: An Introduction.* Cambridge University Press, Cambridge, 1980.

[10] R. Chisholm. Contrary-to-duty imperatives and deontic logic. *Analysis*, 24:33–66, 1963.

[11] J. Forrester. Gentle murder, or the adverbial samaritan. *Journal of Philosophy*, 81:193–196, 1984.

[12] D. Gabbay, J. Horty, X. Parent, R. van der Meyden, and L. van der Torre, editors. *Handbook of Deontic Logic and Normative Systems*, volume 1. College Publications, London, UK, 2013.

[13] L. Goble. The Andersonian reduction and relevant deontic logic. In B. Byson and J. Woods, editors, *New Studies in Exact Philosophy: Logic, Mathematics and Science — Proceedings of the 1999 Conference of the Society of Exact Philosophy*, pages 213–246. Hermes Science Publications, Paris, 1999.

[14] L. Goble. Preference semantics for deontic logics. Part I: Simple models. *Logique & Analyse*, 46(183-184):383–418, 2003.

[15] L. Goble. Prima facie norms, normative conflicts, and dilemmas. In Gabbay et al. [12], pages 241–352.

[16] L. Goble. Models for dyadic deontic logics, 2015. Unpublished (version dated 3 Aug 15).

[17] L. Goble. Notes for the determination of EK, 2016. Unpublished (version dated 10 Jun 2016).

[18] J. Hansen. On relations between Åqvist's deontic system G and van Eck's deontic temporal logic. In P. McNamara and H. Prakken, editors, *Norms, Logics and Information Systems*, pages 127–146. IOS Press, 1999.

[19] J. Hansen. Prioritized conditional imperatives: problems and a new proposal. *Autonomous Agents and Multi-Agent Systems*, 17(1):11–35, 2008.

[20] J. Hansen. Reasoning about permission and obligation. In S. O. Hansson, editor, *David Makinson on Classical Methods for Non-Classical Problems*, pages 287–333. Springer, 2014.

[21] W. Hanson. Semantics for deontic logic. *Logique & Analyse*, 8:177–190, 1965.

[22] B. Hansson. An analysis of some deontic logics. In R. Hilpinen, editor, *Deontic logic: Introductory and Systematic Readings*, pages 121–147. D. Reidel, Dordrecht, 1971. Reprinted from *Noûs* 3, pp. 373-98, 1969.

[23] R. Hilpinen and P. McNamara. Deontic logic: a historical survey and introduction. In Gabbay et al. [12], pages 3–136.

[24] J. Horty. *Agency and Deontic Logic*. Oxford University Press, USA, 2001.

[25] J. Horty. Defaults with priorities. *Journal of Philosophical Logic*, 36(4):367–413, 2007.

[26] J. Horty. *Reasons as Defaults*. Oxford University Press, USA, 2012.

[27] S. Kraus, D. Lehmann, and M. Magidor. Nonmonotonic reasoning, preferential models and cumulative logics. *Artificial Intelligence*, 44(1-2):167–207, 1990.

[28] D. Lehmann and M. Magidor. What does a conditional knowledge base entail? *Artificial Intelligence*, 55(1):1–60, 1992.

[29] D. Lewis. *Counterfactuals*. Blackwells, 1973.

[30] D. Makinson. Five faces of minimality. *Studia Logica*, 52(3):339–380, 1993.

[31] D. Makinson. On a fundamental problem of deontic logic. In P. McNamara and H. Prakken, editors, *Norms, Logics and Information Systems*, pages 29–53. IOS Press, Amsterdam, 1999.

[32] D. Makinson and L. van der Torre. Input/output logics. *Journal of Philosophical Logic*, 29:383–408, 2000.

[33] D. Makinson and L. van der Torre. Constraints for input/output logics. *Journal of Philosophical Logic*, 30(2):155–185, 2001.

[34] D. Makinson and L. van der Torre. Permission from an input/output perspective. *Journal of Philosophical Logic*, 32:391–416, 2003.

[35] J. S. A. Maranhão. A logical architecture for dynamic legal interpretation. In J. Keppens and G. Governatori, editors, *Proceedings of the 16th edition of the International Conference on Artical Intelligence and Law, ICAIL 2017, London, United Kingdom, June 12-16, 2017*, pages 129–138. ACM, 2017.

[36] X. Parent. On the strong completeness of Åqvist's dyadic deontic logic G. In R. van der Meyden and L. van der Torre, editors, *Deontic Logic in Computer Science, 9th International Conference, DEON 2008, Luxembourg, Luxembourg, July 15-18, 2008. Proceedings*, volume 5076 of *Lecture Notes in Computer Science*. Springer, 2008.

[37] X. Parent. Moral particularism in the light of deontic logic. *Artificial Intelligence and Law*, 19(2-3):75–98, 2011.

[38] X. Parent. Maximality vs. optimality in dyadic deontic logic. *Journal of Philosophical Logic*, 43(6):1101–1128, 2014.

[39] X. Parent. Completeness of Åqvist's systems E and F. *Review of Symbolic Logic*, 8(1):164–177, 2015.

[40] X. Parent. Hansson's preference-based semantics for dyadic deontic logic–a survey of axiomatization results. In D. Gabbay,

J. Horty, X. Parent, R. van der Meyden, and L. van der Torre, editors, *Handbook of Deontic Logic and Normative Systems*, volume 2. College Publications, London, UK, (to appear) 2019.

[41] X. Parent, D. Gabbay, and L. van der Torre. Intuitionistic basis for input/output logic. In S. O. Hansson, editor, *David Makinson on Classical Methods for Non-Classical Problems*, volume 3 of *Outstanding Contributions to Logic*, pages 263–286. Springer, Netherlands, 2014.

[42] X. Parent and L. van der Torre. I/O logic. In Gabbay et al. [12], pages 499–544.

[43] X. Parent and L. van der Torre. "Sing and dance!". In F. Cariani, D. Grossi, J. Meheus, and X. Parent, editors, *Deontic Logic and Normative Systems*, volume 8554 of *Lecture Notes in Computer Science*, pages 149–165. Springer International Publishing, 2014.

[44] X. Parent and L. van der Torre. I/O logics with a consistency check. In J. Broersen, C. Condoravdi, S. Nair, and G. Pigozzi, editors, *Deontic Logic and Normative Systems — 14th International Conference, DEON 2018*, pages 285–299. College Publications, 2018.

[45] H. Prakken and M. Sergot. Contrary-to-duty obligations. *Studia Logica*, 57(1):91–115, 1996.

[46] H. Prakken and M. Sergot. Dyadic deontic logic and contrary-to-duty obligations. In D. Nute, editor, *Defeasible Deontic Logic*, pages 223–262. Kluwer, 1997.

[47] A. Sen. Choice functions and revealed preference. *Review of Economic Studies*, 38:307–317, 1971.

[48] W. Spohn. An analysis of Hansson's dyadic deontic logic. *Journal of Philosophical Logic*, 4(2):237–252, 1975.

[49] A. Stolpe. *Norms and Norm-System Dynamics*. PhD thesis, Department of Philosophy, University of Bergen, Norway, 2008.

[50] A. Stolpe. A concept approach to input/output logic. *J. of Applied Logic*, 13(3):239–258, 2015.

[51] L. van der Torre and Y.-H. Tan. Two-phase deontic logic. *Logique & Analyse*, 171-172:411–456, 2000.

[52] J. van Eck. A system of temporally relative modal and deontic predicate logic, and its philosophical applications. *Logique & Analyse*, 25:249–290 and 339–381, 1982.

[53] B. van Fraassen. The logic of conditional obligation. *Journal of Philosophical Logic*, 1(3/4):417–438, 1972.

[54] G. von Wright. Deontic logic. *Mind*, 60:1–15, 1952.